Designing Development

Case Study of an International Education and Outreach Program

Synthesis Lectures on Global Engineering

The Global Engineering Series challenges students, faculty and administrators, and working engineers to cross the borders of countries, and it follows those who do. Engineers and engineering have grown up within countries. The visions engineers have had of themselves, their knowledge, and their service have varied dramatically over time and across territorial spaces. Engineers now follow diasporas of industrial corporations, NGOs, and other transnational employers across the planet. To what extent do engineers carry their countries with them? What are key sites of encounters among engineers and non-engineers across the borders of countries? What is at stake when engineers encounter others who understand their knowledge, objectives, work, and identities differently? What is engineering now for? What are engineers now for?

The Series invites short manuscripts making visible the experiences of engineers and engineering students and faculty across the borders of countries. Possible topics include engineers in and out of countries, physical mobility and travel, virtual mobility and travel, geo-spatial distributions of work, international education, international work environments, transnational identities and identity issues, transnational organizations, research collaborations, global normativities, and encounters among engineers and non-engineers across country borders.

The Series juxtaposes contributions from distinct disciplinary, analytical, and geographical perspectives to encourage readers to look beyond familiar intellectual and geographical boundaries for insight and guidance. Holding paramount the goal of high-quality scholarship, it offers learning resources to engineering students and faculty and working engineers crossing the borders of countries. Its commitment is to help them improve engineering work through critical self-analysis and listening.

Designing Development: Case Study of an International Education and Outreach Program
Aditya Johri and Akshay Sharma
2013

Merging Languages and Engineering: How it Happened
John M. Grandin
2013

What is Global Engineering Education For? The Making of International Educators, Part 2
Gary Lee Downey and Kacey Beddoes
2010

What is Global Engineering Education For? The Making of International Educators, Part 1
Gary Lee Downey and Kacey Beddoes
2010

Designing Development: Case Study of an International Education and Outreach Program

Aditya Johri and Akshay Sharma

www.morganclaypool.com

ISBN: 9781627050036 paperback
ISBN: 9781627050043 ebook

DOI 10.2200/S00470ED1V01Y201301GES004

A Publication in the Morgan & Claypool Publishers series
SYNTHESIS LECTURES ON GLOBAL ENGINEERING

Lecture #4
Series Editor: Gary Downey, *Virginia Tech*
Assistant Editor: Kacey Beddoes, *Purdue*
Series ISSN
Synthesis Lectures on Global Engineering
Print 2160-7664 Electronic 2160-7672

Designing Development

Case Study of an International Education and Outreach Program

Aditya Johri and Akshay Sharma
Virginia Tech

SYNTHESIS LECTURES ON GLOBAL ENGINEERING #4

ABSTRACT

The creation of physical and material infrastructure is the cornerstone of human development; not surprisingly, engineers and designers are often motivated and inspired in their practice to improve the world around them, to make things better for others, and to apply their knowledge for the good of mankind. These aspirations often get translated into engineering and design curricula where students and faculty work on development related projects usually under the category of community or service learning. This book presents an overview of such an education and outreach program designed to empower stakeholders to improve their lives. The project described here was an international multi-institutional undertaking that included academic institutions, non-governmental organizations, and private firms. Within the academic setting, an interdisciplinary set of actors that included engineering and industrial design students and faculty worked on the project. We concretize our work by presenting a design case study that illustrates how different approaches can help guide the works of engineers and designers as they create global infrastructures and localized artifacts. We emphasize the importance of developing long term relationships with organizations on the ground in order to ensure appropriate design as well as successful transfer and long term use of designed artifacts. We discuss the life trajectories of the authors to provide a grounded perspective on what motivated us to undertake this work and shaped our approach with the intention to demonstrate that there are multiple paths toward this goal.

KEYWORDS

engineering, education and outreach, design, global development, ICT for development, sociomateriality, user-centered design, India

Contents

Preface

Through this book, we hope to provide an in-depth look at designing and executing an education and outreach program aimed at benefiting the billions of people living in low resource environments across the world. We present *one* view or perspective that drives a program of this sort and how it can be implemented successfully, relatively speaking, and can be sustained. This case study looks at personal motivations, intellectual influences, on the ground experiences, and other institutional and organizational factors that all play a critical role in an endeavor like this. We hope that readers will be motivated by reading about our experiences and find items of value that they can leverage for their own design. We have also put supporting information online (http://www.id4learning.com; since web links change frequently we encourage readers to search our names to come up with the latest pages).

In Chapter 1 we introduce our education and outreach program by placing it within the larger context of service-learning projects that have become common across university campuses, particularly in the United States but also in other countries across the world.

In Chapter 2 we discuss our personal journeys toward this work and professional opportunities that presented themselves. We first discuss our individual trajectories and follow that up by discussing our collaborative trajectory and the inception of the education and outreach program discussed in this book.

In Chapter 3 we look at some intellectual ideas that have motivated our work. We start by discussing various motivational theories that helped us understand why students and faculty might wish to work on such projects and what motivates us. We then discuss how the interaction of the social and the material or technical plays a critical role in how we perceive our work as building an infrastructure for education and outreach. Finally, we review design frameworks and discuss in depth one framework that has guided our work—capable and convivial design (CCD).

Chapter 4 is the longest chapter in the book and presents an in-depth case study of one design project that we have undertaken over the past couple of years. This project on immunization had its beginning much earlier in another project in which one of the authors participated but has undergone significant design changes in its current iterations. In particular, we look at how Quick Response or QR codes can be integrated with mobile phone applications to improve immunization efficiencies in developing countries, in our case, India. We focus in particular on an urban slum as the context for this design.

In Chapter 5 we present an overview of the different components of our overall program. We discuss a course that we teach and how we integrate service projects in the course. We also briefly discuss our outreach efforts.

In the final chapter, Chapter 6, we outline the different lessons we have learned through our efforts and hope that readers will find it of use in designing their own courses and outreach components.

True to its title of a "lecture" this book is different as we take a more personal approach to discuss development and how to "design" it in an appropriate manner. We document our education and outreach program and through this effort we hope to shape the conversation around the role of scholarly and education practices in development writ large.

Aditya Johri and Akshay Sharma
January 2013

Acknowledgments

We have drawn significantly on the following prior publications for parts of this text:

Johri, A. & Sharma, A. (2012). Learning From Working on Others' Problems: Engaging Students in Long-term Global Projects for Reciprocal Learning. *Proceedings of ASEE Annual Conference 2012*. We draw on this article for sections in chapters 1, 5, and 6.

Johri, A. & Pal, J. (2012). Capable and Convivial Design: A Framework for Designing Information and Communication Technology for Human Development. *Information Technology for Development*, 18(1):61-75. Portions of this work motivate the discussion on the capabilities and conviviality framework discussed in chapter 3.

Johri, A. (2011). The Sociomateriality of Learning Practices and Implications for the Field of Learning Technology. *Research in Learning Technology*, Vol. 19, Issue 3, 207-217. Portions of this work shape the discussion on sociomateriality in chapter 3.

We would like to express our gratitude to Pradaan, our NGO organization on the ground, and the self-help group (SHG) facilitators who worked with us. We are also thankful to the people of Dausa, Rajasthan. Over the past three years we have also had conversations with various other NGOs and we will like to thank them for sharing their experiences with us. We acknowledge the participation of our students in the various design projects. We will especially like to thank Jonathan Ballands[1] and Peter Beegle for their work on the immunization project. Dr. Johri will like to thank the e-Immunization team led by Mr. Rajendra Nimje for raising his awareness of development issues and for laying the groundwork for this project (for details of team members see Chapter 4). He will also like to thank his collaborator Dr. Joyojeet Pal (University of Michigan) for his contribution to the Capable and Convivial Design (CCD) framework. Finally, we will like to express our sincere gratitude to the series editors Gary Downey and Kacey Beddoes for their untiring efforts and patience in helping us improve the manuscript. Any errors, ambiguity, or lack of clarity is the fault of the authors.

This project was partially supported through funds from the Office of International Research, Education and Development at Virginia Tech, Pratt Funding from the College of Engineering, Virginia Tech, and a U.S. National Science Foundation Early Career Award (EEC#0954034) to Dr. Johri.

Aditya Johri and Akshay Sharma
January 2013

[1] http://jonathanballands.me

CHAPTER 1
Introduction

Engineers and designers are often motivated by the desire to have a real world impact through their work. Many engineering faculty and students alike conceive of and implement projects that aim to improve the lives of others through technologies they invent and produce. The desire to make positive changes is present not only among practicing engineers but also among engineering students and faculty. Over the past couple of decades, engineering faculty members across institutions have leveraged this motivation to design courses and experiences for students where they can make a positive impact in the life of others and also learn important engineering and design skills. These experiences have been termed service-learning or community learning. Many service-learning initiatives have been highly successful and have had a significant impact on engineering education. For instance, the Engineering Projects in Community Service (EPICS)[1] program at Purdue University received the Bernard M. Gordon Prize by the National Academy of Engineering for Innovation in Engineering and Technology Education [Coyle et al., 2006]. The EPICS program exemplifies a pioneering educational innovation and has been adopted by dozens of institutions since its inception. In this program, students engage in long-term community service projects in local communities [Coyle et al., 2006, 2005]. Teams are relatively large, representing a small company, and the community organization acts as a client. Students can take the course similar to taking a lab and can enroll multiple times for up to four times. In addition to curricular activities, various societies and groups such Engineers for a Sustainable World (ESW)[2] and Engineers without Borders (EWB)[3] have taken a hold across campuses increasing the opportunities for students to participate in community service activities.

Another model that has emerged is to have project-based courses where the projects that students work on are driven by community needs. In this scenario the course looks similar to traditional courses but there is a significant difference where the course is driven by the project itself. This model makes it easier to engage communities and problems that might not be local to the educational institution, such as global or international development [Bielefeldt et al., 2010, Swan et al., 2010]. In addition to disciplinary engineering faculty, many scholars from Science and Technology Studies (STS) and Engineering Studies have also started to engage with service-learning approaches to examine the ethical and pragmatic difficulties faced in engaging with these projects [Baillie, 2006, Nieusma et al., 2010]. Irrespective of the model adopted, service-learning has become an important part of the engineering curriculum in many institutions since its introduction within engineering

[1] https://engineering.purdue.edu/EPICS
[2] http://www.eswusa.org/; http://en.wikipedia.org/wiki/Engineers_for_a_Sustainable_World
[3] http://www.ewb-usa.org/; http://en.wikipedia.org/wiki/Engineers_Without_Borders

schools in the early 1990s. The newly formed ASEE division "Community Engagement in Engineering Education (CEEE)"[4] is further evidence of its importance and adoption.

1.1 REVIEW OF SERVICE LEARNING LITERATURE

Service-learning as an educational mechanism has found a foothold as a curricular activity since the 1980s and its presence in higher education has increased ever since. The benefits of service-learning projects are well documented in the literature. In a report that summarizes the main findings of research on service-learning, and includes an annotated bibliography of over 100 articles on the topic, Eyler et al. [2001] found that service-learning has a positive effect on students' personal development such as sense of personal efficacy, personal identity, spiritual growth, and moral development and also on their interpersonal development, the ability to work well with others, and leadership and communication skills. Their review further identifies that service-learning has a positive effect on reducing stereotypes and facilitating cultural and racial understanding. Furthermore, service-learning experiences have a positive effect on sense of social responsibility and citizenship skills and students' commitment to service. In terms of learning outcomes, they argue that evidence shows that service-learning has a positive impact on students' academic learning and students or faculty also report that service-learning improves students' ability to apply what they have learned in "the real world." Finally, service-learning participation has an impact on such academic outcomes as demonstrated complexity of understanding, problem analysis, critical thinking, and cognitive development.

In one of the larger studies conducted on service learning [Astin et al., 2000], longitudinal data were collected from 22,236 college undergraduates attending a national sample of baccalaureate-granting colleges and universities. Thirty percent of the students participated in course-based community service (service learning) during college, and an additional 46 percent participated in some other form of community service. The study assessed the impact of service learning and community service on 11 different dependent measures. Service participation showed significant positive effects on all 11 outcome measures: academic performance (GPA, writing skills, critical thinking skills), values (commitment to activism and to promoting racial understanding), self-efficacy, leadership (leadership activities, self-rated leadership ability, interpersonal skills), choice of a service career, and plans to participate in service after college. The study further found that performing service as part of a course (service learning) significantly adds to the benefits associated with community service for all outcomes except interpersonal skills, self-efficacy, and leadership, and benefits associated with course-based service were strongest for the academic outcomes, especially writing skills. Overall, service participation appears to have its strongest effect on the student's decision to pursue a career in a service field. The authors state that the positive effects of service can be explained in part by the fact that participation in service increases the likelihood that students will discuss their experiences with each other and that students will receive emotional support from faculty. Furthermore, both the quantitative and qualitative results suggest that providing students with an opportunity to "process" the service experience with each other is a powerful component of both community service and

[4] http://www.asee.org/member-resources/groups/divisions/ceee/officers

service-learning. Compared to community service, taking a service-learning course is much more likely to generate student-to-student discussions and allow for reflective learning. In terms of motivation, the single most important factor associated with a positive service-learning experience appears to be the student's degree of interest in the subject matter. Subject matter interest is an especially important determinant of the extent to which (a) the service experience enhances understanding of the "academic" course material, and (b) the service is viewed as a learning experience. These findings provide strong support for the notion that service learning should be included in the student's major field. The second most significant factor in a positive service-learning experience is whether the professor encourages class discussion. The qualitative part of the data analysis suggests that service learning is effective in part because it facilitates four types of outcomes: an increased sense of personal efficacy, an increased awareness of the world, an increased awareness of one's personal values, and increased engagement in the classroom experience. Both qualitative and quantitative results underscore, once again, the power of reflection as a means of connecting the service experience to the academic course material. The primary forms of reflection used were discussions among students, discussions with professors, and written reflection in the form of journals and papers.

1.2 INTERNATIONAL SERVICE LEARNING

Moving beyond engagement with local communities and projects, an international or global flavor is also evident in many service-learning projects. In many cases this interest has emerged out of engineering faculty members' research efforts. Many faculty members in civil and environmental engineering, mechanical engineering, and other disciplines, are involved with research on global climate change, sustainability, and other global challenges. Many researchers and educators are driven by a prosocial motivation to help others in less fortunate circumstances and by the realization that long term sustained development requires significant physical and material infrastructure as evidenced by the visible symbols of development such as dams and roads, but also by the power and sanitation infrastructure. This belief in the positive outcomes of engineered environments is confirmed through reports such as U.N. Millennium Development Goals[5] material aspects of poverty reduction, health issues, financial literacy, and so on. For engineers and designers, there is also an increased awareness that there is a lot to be learned by working on problems in different contexts as they teach you about designing in constrained environments and lead to innovative ideas and concepts that are applicable universally.

1.3 GOALS OF THE BOOK

In this book we report on an educational and outreach program aimed at designing and implementing information and communication technologies in support of development efforts. The work we report here has all been done within the context of India. The overall program consisted of multiple elements: (1) research on existing and potential solutions, (2) design of new solutions as part of a

[5] http://www.un.org/millenniumgoals/

course and independent study credits, and (3) testing and integration of solutions in partnership with stakeholders. The motivation for this work came from our personal experiences in conjunction with professional opportunities that became available to us. In our efforts, we were driven by several intellectual perspectives that all emphasized a deep understanding of the design context and the role of technology within it. In this book, we start by outlining how we came together to design and implement the program and what were our guiding principles. We then present an in-depth case study of one design project—the use of information technology for improving immunization efficiency. This case study brings together both elements of our program—education and outreach. We then focus on how we integrated our work within curriculum practices through the design of a course. We believe that for faculty attempting to undertake this work within the context of educational institutions this chapter will be useful. We conclude by outlining a few lessons we have learned and emphasize the importance of developing a deep understanding of the context within which you plan to undertake the work.

CHAPTER 2

Development of the Program: Personal Trajectories Meet Professional Opportunities

In this chapter, we provide an overview of our personal journeys[1] into development work and narrate how those journeys merged with professional opportunities that brought us together and resulted in the creation of our education and outreach program. Our personal journeys or trajectories outline our experiences and are both similar in terms of their focus but also different as we approach our work from diverse perspectives. One of us, Johri, focuses more on the professional side whereas Sharma has a more personal approach. We present these different journeys to show that folks with diverse perspectives and approaches can form fruitful partnerships.

2.1 TRAJECTORY INTO GLOBAL ENGINEERING DESIGN AND EDUCATION—ADITYA JOHRI

My interest in development issues started in earnest when I participated in a social entrepreneurship competition during my doctoral studies at Stanford University. The team effort was led by an administrative officer from the Indian Civil Services who was on sabbatical at Stanford. As part of this project we worked on designing an immunization system to assist health workers in rural India with their data management needs. Prior to this engagement, although I had done volunteer work for several non-governmental organizations (NGOs) involved with developmental work in India, such as Child Rights and You (CRY),[2] and ASHA for Education,[3] I had not been directly involved with design of a product. The "e-Immunization" project, as we called it, was a learning experience on many fronts. It was my first time interacting closely with someone who had been an integral part of the Indian Civil Services[4] and it was interesting to learn how they viewed things from the inside. The contextual knowledge provided by our mentor was invaluable. For instance, immunization rates are extremely low but that is not the case for all immunizations. The rates for vaccinations for polio, for instance, which were quite low, have been dramatically increased. Polio requires one dose any time

[1]The inspiration for this approach comes from similar trajectories included in this edited volume: Downey, G. & Beddoes, K. (eds), *What is Global Engineering Education For?*, pp. 415-432, Vol. 1.
[2]http://en.wikipedia.org/wiki/Child_Rights_and_You
[3]http://www.ashanet.org/
[4]http://en.wikipedia.org/wiki/Civil_Services_of_India

before the age of 3 and therefore if you can create enough momentum for parents to get their children vaccinated once, it is sufficient. Many other vaccinations though have a more complicated delivery mechanism. They require several doses over a specific time period and often at specific intervals. Now imagine a health worker in a rural part of India who is responsible for immunization across several villages. The information tracking required for this job is immense. First of all, you have to be cognizant of expectant mothers and when babies are due for delivery. This is not an easy task as most deliveries are still performed by midwives without reporting to a central office. Then, for each new child born, information has to be managed when they were vaccinated and when the upcoming vaccinations are to be given. Parent involvement is often minimal given high illiteracy rates as well as the lack of time on the part of the parents who are busy either with earning or working in the fields or at home. This complicated context is the reality of the health worker. From the perspective of the companies that manufacture the vaccines, this is for them a money losing proposition. They can devote these resources for manufacturing medicines that are a lot more profitable. Therefore, they have asked Global Alliance for Vaccines and Immunization (GAVI)[5] to ensure that information about what is needed is well maintained. Storage of vaccines is another issue that has to be dealt with. Overall, this project highlighted the complexity of development work and the different actors that play a role in the larger ecosystem. As we worked on the project we realized that there has to be a way to address all these issues. The project, in retrospect, was a technologically determinist approach. The infrastructure that already existed in that state was a key factor in motivating us to look at the mobile/handheld issue. Hardware issues came into play as cost and local manufacturing was important. Overall, we ended up doing a good job of balancing social needs and technology affordances. The solution we came up with worked as follows. Health workers were given a handheld device that had a smartcard and a small printer in addition to a keyboard. We did a lot of research to select the device and to ensure that it was compatible with the local environment; we worked with a small regional firm that made devices for use in the local government. The device was sturdy and could be repaired easily. We were able to leverage the mobile network already available in the state for data transmission from rural areas. We were able to build significant redundancy in the system—a crucial requirement for technology use in developing areas. Finally, the paper print out although not useful from the data perspective allowed us to leave something with the parents that made them get more involved in their child's health. The parents felt empowered.

While I was working on the e-Immunization project, I also had the opportunity to engage with another project called BookBox.[6] This was a really interesting project that aligned with a lot of interests that I had in the use of media for social good and the education-entertainment paradigm. The project was based on a very simple idea—same text subtitling for literacy. In other words, viewers saw a video but were also provided subtitling in the same language to help them develop their literacy skills. The project targeted an illiterate population that was fluent in spoken language but could not read or write—a situation common in many emerging economies. The project was conceived by

[5] `http://www.gavialliance.org/forum/`; `http://en.wikipedia.org/wiki/GAVI_Alliance`
[6] `http://www.bookbox.com/`

Dr. Brij Kothari and he had successfully tested the idea where movie songs shown on TV were subtitled and similar to Karaoke had a line moving across the text syncing the spoken and written words. As a next step Dr. Kothari was working on taking this idea and using it to produce electronic text books. My very short interaction was a great learning experience as it taught me about having small but useful ideas and the importance of scaling them. It also made me cognizant of the time and effort required to actually scale any great idea—lesson that was continuously present given the entrepreneurial nature of Silicon Valley.

During one of my trips to India during this time, I had the opportunity to visit Microsoft Research (MSR) Labs in Bangalore.[7] MSR India is one of the leading institutions conducting research on the role of computing in development. During my trip I was fortunate to talk with research scientists and graduate and undergraduate students who were all interested in issues of development. This visit gave me a very different perspective on a whole different population of people who were in this area—researchers. Most of the projects that were being undertaken at MSR were prototypes that demonstrated the usefulness of ideas. One strong component of their research was the amount of time researchers spent in the field understanding the design context. They also spent considerable time in building relationships with communities and NGOs in order to test and if possible implement the products they designed. The research lab also demonstrated the international interest in this topic. Although the majority of the researchers were of Indian origin there were many researchers from other nations.

The researchers at MSR were part of a new field of scholarship called information and communication technologies for development (ICTD or ICT4D).[8] In this period, I was also in conversation with a group of colleagues at Berkeley working in the area of ICTD. The field had brought together engineers and social scientists to work on solutions for the developing world. Although these kinds of partnerships are not new, the significant advances and adoption of IT across the world has really brought people together in search for solutions to improve humanity. This work, of course, has its critics who argue that a technological determinist attitude pervades in the field and therefore the majority of designed solutions fail.

The ICTD discipline grew around the dot-com boom, and Silicon Valley at that time had just recuperated from the shock of the dot-com bust. Firms that survived, and the millionaires that emerged, were looking for ways to "do good." As a result, there were a number of start-up companies that were exploring how technology could be used for these purposes. One of the firms that was just formed and coming up was KIVA.[9] In a presentation I attended by Premal Shah who was leading the product design at KIVA I was amazed by not only the uptake of their product by those in developed countries but some of the uses their system was being put to by users in Africa—the unintended consequences are often the more relevant and interesting issues in technology and development work.

[7] http://research.microsoft.com/en-us/labs/india/default.aspx
[8] http://en.wikipedia.org/wiki/Information_and_communication_technologies_for_development
[9] http://www.kiva.org/

2.1.1 GOING BACK IN TIME

The immunization project was a firsthand learning experience and a trigger for future design work in the area, but what made it important to focus on these issues? The answer to that, on reflection, is the change that technology—communication media—in particular, can bring about for development. The role of technology in development was evident to me first hand as I grew up in a land grant university in India, at Pantnagar,[10] that was the "Birthplace of the Green Revolution." I grew up hearing stories about the lack of food grains in India and the dependency of India on low quality food grains donated by nations such as the US. India overcame this problem thanks largely to the introduction of hybrid crops, similar to those introduced in Mexico by Norman Borlaug.[11] At the university where I grew up the same research practices were used to modify other crops, particularly rice which is a staple of Indian diet. Slowly this revolution spread and crops such as soybean that were not even found in the country became popular. The cycle of crop sowing and reaping changed from one to two per year and then three with a short cycle of legumes in between. For a nation of 1.2 billion people self-sufficiency in food is a significant achievement. These practices, of course, did not come without long term environmental problems such as depletion of soil and significant lowering of the water table. In addition to the crop though, another significant movement was the movement of mass communication that accompanied changes in agricultural practices. On reflection and having spent time in a land grant university in the US it is easy to see how this was adopted from the US land-grant model.[12] Yet, communication through radio was a significant part at the community level. To institutionalize the practices messages that were related to farmers' practices was critical. This piqued my interest in social side of technological progress. How does human behavior change? What is the relationship between the medium and the message? At this time I came across Marshall McLuhan[13] and although I did not understand his work much, his primary message—that the medium of communication changes social functioning more than the content—was clear to me. It was also easy to understand the message as slowly, and right in front of my eyes, I could see the implications of his words in action. It started with the introduction of cable TV in our small town and the significant shift in leisure practices that accompanied it. As I started college and then my job, the use of the Internet was changing the practices of work and scholarship. And now, with the advent of mobile phones, the overall fabric of interaction among the masses has changed fundamentally. This, of course, does not imply equality, rise in health, and other indicators of "development" but that the medium is the message is clear.

This is also indicative of the design and materiality approach that I take in my work on development. Through my years in academia I have been fortunate to work with scholars in different traditions, including sociologists, psychologists, and economists, who all have their unique approach toward addressing issues of development. The overall picture presented by economists and those who work with large datasets is a useful way to understand the overall picture and indicators but is

[10]http://en.wikipedia.org/wiki/Pantnagar
[11]http://en.wikipedia.org/wiki/Norman_Borlaug
[12]http://en.wikipedia.org/wiki/Land-grant_university
[13]http://en.wikipedia.org/wiki/Marshall_McLuhan

not contextual enough to address real world problems on the ground. Working on the field has also attuned me to the challenges that present themselves. As in any socio-technical project, working on development issues is a tough challenge as the technology, materiality, is just one part of the puzzle. Over the past couple of years I have worked on research projects examining large-scale deployment of IT infrastructure and design of devices for the purpose of bringing about large scale change within India. Given the technological deterministic discourse that is prominent in India given its success in providing IT services, there is a significant push by the government to utilize IT as a way to increase efficiency of governance. This objective and its implementation can be seen in examples as diverse as the use of electronic voting machines as well as recent implementations of online service for payment of utility bills online.

2.1.2 PERSONAL TRAJECTORY MEETS PROFESSIONAL OPPORTUNITIES

My interests in topics related to development, particularly ICT for development, were further nurtured through professional opportunities that became available to me when I joined Virginia Tech (VT) in 2007. Although VT does not have an official research group working on these issues, many faculty and students are interested in the topic. For instance, I was able to collaborate with faculty in the School of Education [Evans et al., 2008] as well as scholars in the Center for Human-Computer Interaction (CHCI)[14] and the program in Science and Technology Studies (STS).[15] Furthermore, I was able to extend my collaborative network to scholars beyond VT and work with researchers from all over the world. More importantly, I was able to garner research funds through different initiatives at VT and through other funding agencies to undertake this work. Personally, in addition to education and outreach, I was also involved with ethnographic and qualitative research in how ICT innovations were adopted in developing regions, and publishing this work formed a component of my overall research agenda. Two such projects included a field study of a job guarantee scheme for rural India called NREGA[16] (Figure 2.1) and a project that aims to provide biometric based identification to all residents of India. This project, called Aadhaar,[17] is one of the largest information technology projects in terms of the size of implementation ever attempted. It is also a unique and successful example of public-private partnership. While I was continuing on my personal trajectory I met Prof. Sharma and subsequently we decided to join hands to undertake collaborative work including visits to self-help group (SHG) sites engaged in microfinance (Figure 2.2). Our collaboration is discussed in depth later.

[14] http://www.hci.vt.edu/

[15] http://www.sts.vt.edu/

[16] http://nrega.nic.in/netnrega/home.aspx; http://en.wikipedia.org/wiki/Mahatma_Gandhi_National_Rural_Employment_Guarantee_Act

[17] http://uidai.gov.in/; http://uidai.gov.in/index.php?option=com_content&view=article&id=57&Itemid=105; http://en.wikipedia.org/wiki/Unique_Identification_Authority_of_India

Figure 2.1: A farmer near his well—the project was funded by the NREGA scheme in India (Photo by Aditya Johri).

Figure 2.2: Aditya Johri leading a feedback session with a self-help group (SHG) in Rajasthan, India.

2.2 TRAJECTORY TOWARD DESIGNING PROJECTS RELATED TO DEVELOPMENT—AKSHAY SHARMA

2.2.1 GROWING UP IN INDIA

I grew up in a small town in India. Born to middle-class parents, I lived in a township that was part of a manufacturing facility that made instrumentation for industries in India. My father was a draftsman, one of the 30,000 that were employed by the company at that time. The town was filled with educated people from all over the country who had found productive work in this city. I grew up in a very cosmopolitan environment. To understand the diversity, one has to realize that India has 25 official languages with over 2,500 dialects. We had a good number of neighbors who did not speak the same language or who practiced a different religion. The unifying factors were the identical dwelling units we all lived in, the same school all the kids went to, and the same park all the kids played in. Even though there was uniformity in the different aspects of our lives, I soon realized that some of my friends lived in bigger quarters, some of the parents drove newer vehicles, and some kids went to a different, supposedly better, school. For a long time I didn't understand why my father was always pestering me to study hard, and I did not listen to him. Later, when I understood these differences, I saw that he had not wanted his children to feel that what he had accomplished was in any way less than his colleagues. He was a very disciplined, hardworking, and productive person. My father made about $5 per month (250 rupees) and supported a family of five after giving half of his salary to my grandfather. It is still hard for me to fathom how he managed the family finances, but I am here writing this narrative because of my parents' careful resource management. I do not recall my parents ever wasting anything. While my father worked in the factory, my mom took care of us. In the afternoons, she sewed clothes for the ladies in the township. She was sought after for her skills as she was able to achieve the right fitting using a set of sewing patterns that my grandmother gave her on her wedding. I remember watching her using those old newspaper patterns to mark the fabric with blue tailor's chalk and the rhythmic sound of her sewing machine, which had been assembled by my grandfather, who used to make parts for sewing machines in his little machine shop in the village where my mother grew up. She turned the small pieces of leftover fabric into doll clothes for my elder sister and her friends.

Every single aspect of my life has been shaped by the first ten years I spent in that township. The first tricycle I got was a hand-me-down; it was refreshed and repainted by my father and reupholstered by my mom. My father made my first cricket bat out of wooden slats from a mango crate, and my mom created a custom grip using old sari fabric. When I was around four or five years old, my father made an air-cooler for our apartment—he bought the different parts, made the sheet metal body on the terrace, and assembled the whole thing before installing it in the window. He was a superman in my eyes. It did not take me too long to become his self-appointed helper. When he would take apart his Vespa scooter to rebuild the engine and paint the body, I would clean the small parts in the carburetor using a toothbrush. When he would restring his badminton racquet, I would watch him intently as he used a needle to get the correct tension in the strings. Another very important aspect of my upbringing was the different environment in our house. In those days, even now in most cases, women were supposed to take care of the household and men were supposed to take care of the outdoor chores. But in our house, my father loved cooking, and he would work with my mom in the kitchen on a regular basis. Another aspect of growing up in those days was the disparity between girls and boys. It was unheard of for boys to do household chores, like helping in the kitchen or sweeping the floor with a broom or mopping the floors clean. However, my elder brother and I were put to work from very early on helping with household chores. I learned about equality, respect, and, above all, an absolute commitment to honesty. The results of this approach were very clear to me by the time I was in the seventh grade. My father was the first among his colleagues to buy a house of his own and move out of the company-provided accommodations. It was a big deal, and he did it simply by saving every single penny he could. One of the last conversations I remember having with my father, when I was a freshman in architecture school, was related to success and prosperity. He reminded me of the value of patience and hard work and about having respect for the people who do the seemingly menial jobs. He said, "Just because you will be an architect does not mean that the work of a mason becomes less important when you show up on the site. How much a person makes should not dictate your behavior toward him." Sincere, honest, loving, and adamant with his principles is how I remember my father, who died when I was still a freshman in college. I wish he could see the impact of his teachings on my life today.

The school I went to was typical for most students in India. Prayer meeting in the morning, with the whole school singing together, followed by the daily news reading by senior students (these were the pre-TV, pre-electronic media days). My book bag was big, but that was never the focus of going to school. In fact, I do not remember anything about my early schooling days. I was happy-go-lucky playing in the street with old scooter tires and playing cricket in the playground, and I always managed to "fall asleep" before 9:00p.m., which was when my father came home from playing badminton. The idea was simple: If I was asleep, I couldn't be made to study. I now regret not having been more diligent with school, and I wish the teachers had been more encouraging and that learning had not been such a burden. One advantage of growing up in a regimented community is that everything was defined for me: the school I went to and the places I visited. My siblings and I never really felt the need to explore anything else. It was not until later when I moved to a bigger city for my high school education that I realized there was a lot more to the world. Staying with relatives made me realize how easy it was to live with my parents: like how my parents understood my needs even before I said something versus having to explain why I needed something to finish my work. Moving out of the comforts of family life made me realize the importance of taking responsibility for my actions. I could no longer just assume that my father would fix anything that I broke or messed up.

I never questioned my teachers or the way we were taught in school until I was in sixth grade. At the school I attended the instruction was in Hindi, my native tongue, but English was used for the written exams. But I had a desire to be able to express myself in English. One day I gathered up enough courage to ask a question in English. The teacher asked me not to try that again. Although I wanted to learn English because my father said it was important, it took me another 13 years to try again. That was my first interaction with social hierarchy, and it left a big impact on my psyche. I suddenly realized that some of the kids lived in bigger quarters because their fathers had higher posts. That's also why they went to the special school outside the township, a true English medium school where the instruction was in English. Although I had a modest upbringing, my parents made sure that my siblings and I never felt a lack of the necessities of life. Until that day in sixth grade, I never realized that there was any difference between the school I went to and the English medium school others attended. When I would ask my father why some kids went to a different school and why I couldn't go there too, he always explained that it takes an hour to travel there and that it was not better than the school I was going to. Much later I realized the true reason: how much he could afford to spend on his kids' education. Another big turning point in my life came in high school. We were required to pick a subject, science-math, commerce, or humanities. I wanted to do science-math. The school in my hometown did not offer that subject, so I ended up moving to a bigger city and was able to secure admission in one of the best missionary schools in Rajasthan. I did not do well for the first semester and was in a state of shock. Eventually I realized that my parents were making a big sacrifice for my education, and I improved my grades.

I still had no clue what I wanted to do and why. My next door neighbor was an architect, and I liked building things, so I decided to give that a try. To get into Architecture College, I had to appear for an entrance examination where 36 students out of 13,000 would be selected. I was number 35. I felt elated and excited that I had finally found what I was really interested in. I enjoyed every single moment of my education at the School of Planning and Architecture in New Delhi. I took a conventional route in school and an unconventional one after it. Right after graduation, instead of joining an architectural practice, I joined a young multidisciplinary design firm and worked with them for three years. One of the projects I worked on was a year-long prelaunch event for a new car by Mitsubishi in India. We designed and supervised the fabrication of a mobile exhibit and took it on the road for 11 months. I was in charge of the show for the duration of the project and had a team of 12 very hardworking men. I was the youngest but had to manage a team of guys who were all much older than I was. We travelled 25,000 miles, visited 17 cities, and displayed new cars at 55 locations. This was also my first experience with folks who live at the bottom of the pyramid, the almost 4 billion people worldwide who live below the poverty line. The team mostly came from villages; they were in search of a better life in New Delhi. They often ended up as security guards: very little pay for hours of physical duty. When we offered them another opportunity, which required lot more work but offered much better remuneration, we had twelve individuals who were willing to go above and beyond their responsibility just to get a chance at a better life. My father always said that it's not how much you make but how much you save. I saw the practical application of that philosophy. The men were getting paid just enough to afford a place to sleep when we were visiting different cities, but all of them decided to sleep in the trucks to save as much as they could and cooked their own meals. I am still in touch with a few of them. In retrospect, those eleven months shaped my ideology and laid the foundation for what I am doing today.

2.2.2 COMING TO THE US

I came to United States as a graduate student. Just going through the admission process and securing a student visa seemed like a huge accomplishment. I landed in Tempe, Arizona, on July 29, 1999. I was not really sure what to expect. One of my biggest surprises was my first meeting with the industrial design program chair at Arizona State University. It was shocking to talk to my professor in what seemed like a casual conversation with a friend. I was used to a different kind of structure in India where the professor was the one dictating the terms and the students were never supposed to question anything. The concept of having an open conversation with my professor to figure out the course of my graduate program was alien to me. The two years at ASU went by fast, and before I knew it I was in San Diego working as a staff industrial designer at a manufacturing firm. A turn of events led me to Blacksburg, Virginia, working for the same company. Over time I got an opportunity to teach part time at the Virginia Tech Industrial Design Program starting in 2003. I enjoyed that experience and decided to try teaching for a little while longer. In those two short years I was able to work on a series of collaborative projects with some of the leading names in Autonomous technologies and felt that I had contributed in a constructive manner. The initial success helped me decide to go into education full time.

2.2.3 PERSONAL TRAJECTORY MEETS PROFESSIONAL OPPORTUNITIES

In 2007 I was offered a tenure-track position; that was a point of no return for me. The first two years were spent in building bridges and figuring out what I really wanted to do. Sketching based ideation has been a focus area in my research, and I spent the initial years developing a new methodology to teach ideation to industrial design students. Soon I understood the basics of teaching the subject, and it was not very long before I started exploring the deeper meaning of how and why we draw. A simple act of making marks on a piece of paper can be made interesting if it is linked to the cognitive process. I further researched this area and created a collaborative project with a faculty member who taught 3D modeling software in the School of Visual Arts. For this project, we taught 3D modeling the way sketching is taught to the students in industrial design. The positive results of this project were enough to demonstrate the potential of constructive collaboration with experts in fields other than mine, the beginning of a rewarding journey in multidisciplinary collaboration. The initial years were tough; I did not have a clear focus. It got to a point where I started thinking about quitting. The pressure to perform and get published while still teaching five days a week was too much. It was during that phase that I was invited to present my research on using sketching techniques to teach complex 3D modeling applications at an international design educators' conference in Singapore. The theme of the conference was "our world in 2050," and it had a plethora of big-name speakers from the world of design. The talks focused on how technology will make it possible for us to have flexible automobiles and how technology will help us counter the rising sea levels.

The conference also coincided with the arrival of my son. All of a sudden the meaning of life changed for me. I started thinking a lot more about the state of the world and the kind of society my son would grow up in. The talks at the conference unsettled me. I was not sure if I agreed with any of the approaches being presented. Here we were talking about the crises that we were in and proposing a series of solutions, but no one was talking about the majority of the human population who had very little to do with the crises—those who live at the bottom of the pyramid. When you are surviving on less than $2 a day, you cannot have a huge impact on the consumption of resources. That conference shook me to the core. I started questioning my purpose in education and in life. My memories from childhood, growing up in a small town, and working with the 12 wonderful team members in the early years of my professional career all came back. I knew I wanted to do something to change the status quo. Being part of an excellent program at Virginia Tech also helped since I knew that I had students who would be willing to be a part of such initiatives. Before I could explore a solution in the university setting with students, I needed some answers and some insights for my own clarity. I initiated a personal exploration, which eventually took me to India, that I was not even sure was going to lead anywhere. I wanted to learn if it was actually possible to change the status quo. I started to research programs that have worked to alleviate poverty. I looked at charitable organizations and also studied the amount of money that is donated every year. I realized that the problem is not the funding or the will to change things. It has to do with the elitist attitude the developed world has toward the developing world.

This was also my opportunity to get back to my roots. I traveled to India, and every morning I would get up early, have a traditional Indian breakfast with chai, kick start my trusty but not very reliable 1971 Enfield Bullet motorcycle,[18] and ride for an hour or two to the village where I was conducting my research. I spent a couple of weeks on the preliminary research where I would walk up to an unknown group of women in a village that I had never been to and engage them in a conversation about life—a surreal experience. It was those moments that laid the foundation for my future initiatives. I met with people from a diverse cross section of the rural Indian society, from young mothers to grandmas. The common link was poverty. This study also affirmed a suspicion I had had from the very beginning. Men usually have it pretty good; it does not matter if they are poor or not. One way or other the women take care of the family and in general do not complain. Those few weeks of intense engagement brought back memories from my childhood—as a boy every summer I used to visit the village that my mom grew up in—but with a layer of new perspective. I now understood why my grandma got up at 4:00a.m. to grind the grain in a hand mill, milk the cows, and have a meal ready for the whole family before they left home for the day. The daily routine of the women made sense to me now. It was designed to facilitate the smooth running of the day for everyone else. On my visit to the villages I realized that nothing had changed. The bulk of work is still being done by women, and they do it happily.

Besides witnessing the women's day-to-day activities, I also attended many self-help group (SHG)[19] microfinancing meetings [Khavul, 2010, Kumaran, 1997]. A typical SHG meeting starts at a predefined place with the arrival of the group accountant and some of the women (Figure 2.3). Since most of the members live in the neighborhood, a quick yell or sending kids to call the members who are missing initiates the proceedings. Until I actually attended these meetings, I did not understand the participants' level of sincerity and commitment to the group. We are talking about 15 to 20 women who have never stepped foot outside their predefined roles in life: taking care of kids, cooking, and cleaning, their entire life. All of a sudden they are tasked with the additional responsibility of not only saving an additional 20 rupees every two weeks but also attending the mandatory meetings. Reading about micro-financing will never relay how intermingled the personal lives of members are with the SHG meetings. In most cases during meetings the members are looking over their shoulders to see what is going on in their yards or where their kids are playing. The biggest revelation for me was how participating in a micro-financing group had empowered these women, where they are now talking about issues that are important to them and collectively making decisions. For instance, in one of the meetings I attended the women discussed how the local school headmaster was not providing midday meals to the students who were supposed to get them. The women all decided to go to the

[18] `http://en.wikipedia.org/wiki/Royal_Enfield_Bullet`
[19] `http://en.wikipedia.org/wiki/Self-help_group_(finance)`

Figure 2.3: A self-help group (SHG) working with an early prototype of an accounting tool in Rajasthan, India.

school after the meeting. They asked the headmaster to do his job or face a complaint to the Block Development Officer[20] (also see Figure 2.4) in charge of the midday meal program. The issue was quickly resolved.

This ethnographic research provided me with numerous such insights and a greater understanding of the true impact of micro-financing. Even though the primary driver for the members is savings, the true benefits are the sense of empowerment through collective decision making and their care for each other. The whole concept of collective self-determination was evident in the meetings. If a member who was supposed to repay the loan experiences a hardship, the group can decide to give her more time or extend another loan. The evidence of empathy for each other in the process and its impact on the productivity of the SHGs is very inspiring. No other financial

[20] http://en.wikipedia.org/wiki/Block_(country_subdivision)

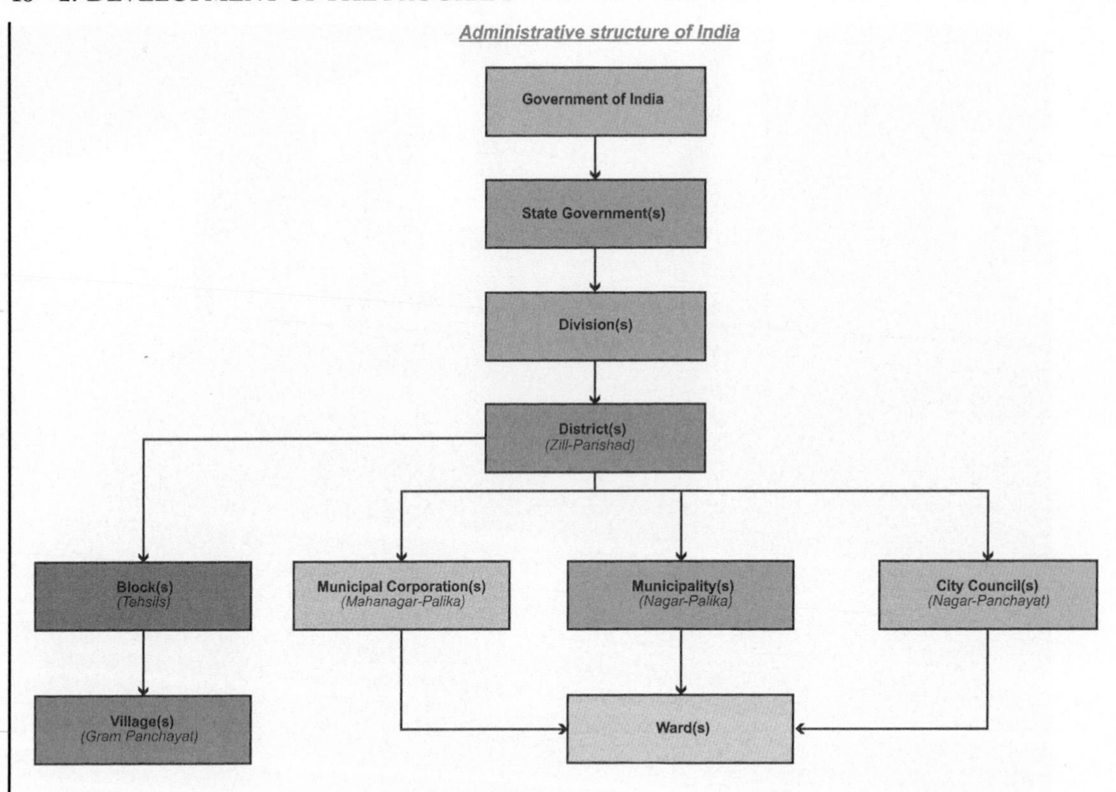

Figure 2.4: Administrative Structure in India.[21]

service provider can claim to have a loan repayment rate of 99.7 % with over 78 million members worldwide. The other part of my ethnographic study took me to Barefoot College[22] in the village of Tilonia in Rajasthan, India. It is a social science research institute founded in 1975 to create a model of sustainable living in rural areas. Its primary goal has been to provide better opportunities for the local population through education and skill development. I grew up not far from this village, but it took me 40 years to pay my first visit. Today it takes only an hour on the newly built expressway to get to the exit, but the road after that is reminiscent of the reality in most of the developing world, taking about 40 minutes to cover the last mile of the journey. The campus itself is a great example of participatory architecture, where the local community was involved at every stage of the design process. They designed it, built it, and now they use it. It all seems like a dream with solar panels everywhere, work areas with rural folks teaching each other, and a community kitchen where the afternoon meal is cooked using a solar cooker. It is a living lab of extraordinary nature.

[22]http://www.barefootcollege.org/

Figure 2.5: Solar cooker at Barefoot College, Rajasthan, India.

The campus is run on solar energy generated by a solar panel system manufactured by Barefoot Solar Engineers. On another part of the campus Barefoot Engineers are busy manufacturing solar cookers, which are used not only on the campus but also in day-care centers in the region. These cookers are complex devices that require an in-depth understanding of solar trajectory to achieve proper setup and optimal performance. The engineers can barely read and write, but they can manufacture, install, and maintain these cookers at the highest professional levels (Figures 2.2.3 and 2.2.3). The workers are mostly women, and they also train new students from Afghanistan, Sub-Saharan African countries, and countries in the Indian subcontinent. The Barefoot College Village Dentist is a program run by two semi-literate women who received six months of training from a visiting dentist from Germany. Today the clinic provides dental services to approximately 30,000 residents in the area. It also runs awareness clinics for local schools and communities. Though the treatment is very affordable, it is free for those who cannot pay for it. I visited Barefoot College to get a better idea of the design opportunities available there. Education opens doors and provides opportunities, but it is also difficult to get a foothold in the challenging world of good schools and institutions, especially for people at the bottom of the pyramid. It is not that the poor do not value education, but rather that other more pressing needs dictate their day-to-day lives: how to feed their families, how to earn a livelihood, and how not to lose whatever little they have. A very surprising discovery was the level of happiness I saw in the villages, despite the poverty. I felt that I had discovered my inner calling. It was as if I had been waiting all these years to show up and be embraced by these people who were used to organizations coming in and suggesting solutions but that never sat down and talked with them about what they really wanted.

Figure 2.6: Prof. Sharma (in the center, wearing a black jacket) and Virginia Tech students inspecting a solar cooker at Barefoot College, Rajasthan, India.

2.3 COMMON PATH FORWARD: INCEPTION AND DEVELOPMENT OF THE PROGRAM

The idea for the program discussed in this book came at a chance meeting between the authors. Johri was organizing a workshop at a conference and Sharma came across that call for papers and our collaboration began from there. Although both the authors had different prior experiences—Johri had previously engaged with a project on immunization (e-Immunization) and conducted a study of rural employment, while Sharma was working on a project on microfinance (SHG)—there was enough common ground to motivate them to plan joint projects and a trip to India. Through this trip new partnerships were developed with NGOs and firms in India and existing partnerships were strengthened. The preliminary collaborative work was presented at a workshop and subsequently an opportunity emerged to engage students for summer research experiences. Students from our home institutions as well as other external institutions worked on projects over the summer. The summer projects were further solidified in a course offered in fall 2011 in which 20 students from engineering and industrial design worked together on projects. A faculty from computer science,[23] who does research on technology and social development, also joined the team. Together the faculty team had experiences in a range of settings and projects including Asia and Africa and areas of work such as health, education and literacy, energy, and others.

Throughout our joint effort, partnerships on the ground were deemed important and cultivated through regular interaction with collaborators. Inter and cross-disciplinary work was emphasized as real world problem solving requires different viewpoints and expertise. So far, over 30 students have participated in the project. The course continues to be offered and is currently going through the governance process to be turned into a regular offering in both engineering and industrial design. The details of the course and our outreach efforts are discussed in depth in Chapter 5. See Figure 2.7 for a timeline of our collaborative efforts.

[23]Dr. Susan Wyche, now at Michigan State University, was a great project partner and brought deep expertise in human-computer interaction and field work in Kenya. For more information see: `http://www.susanwyche.com/`

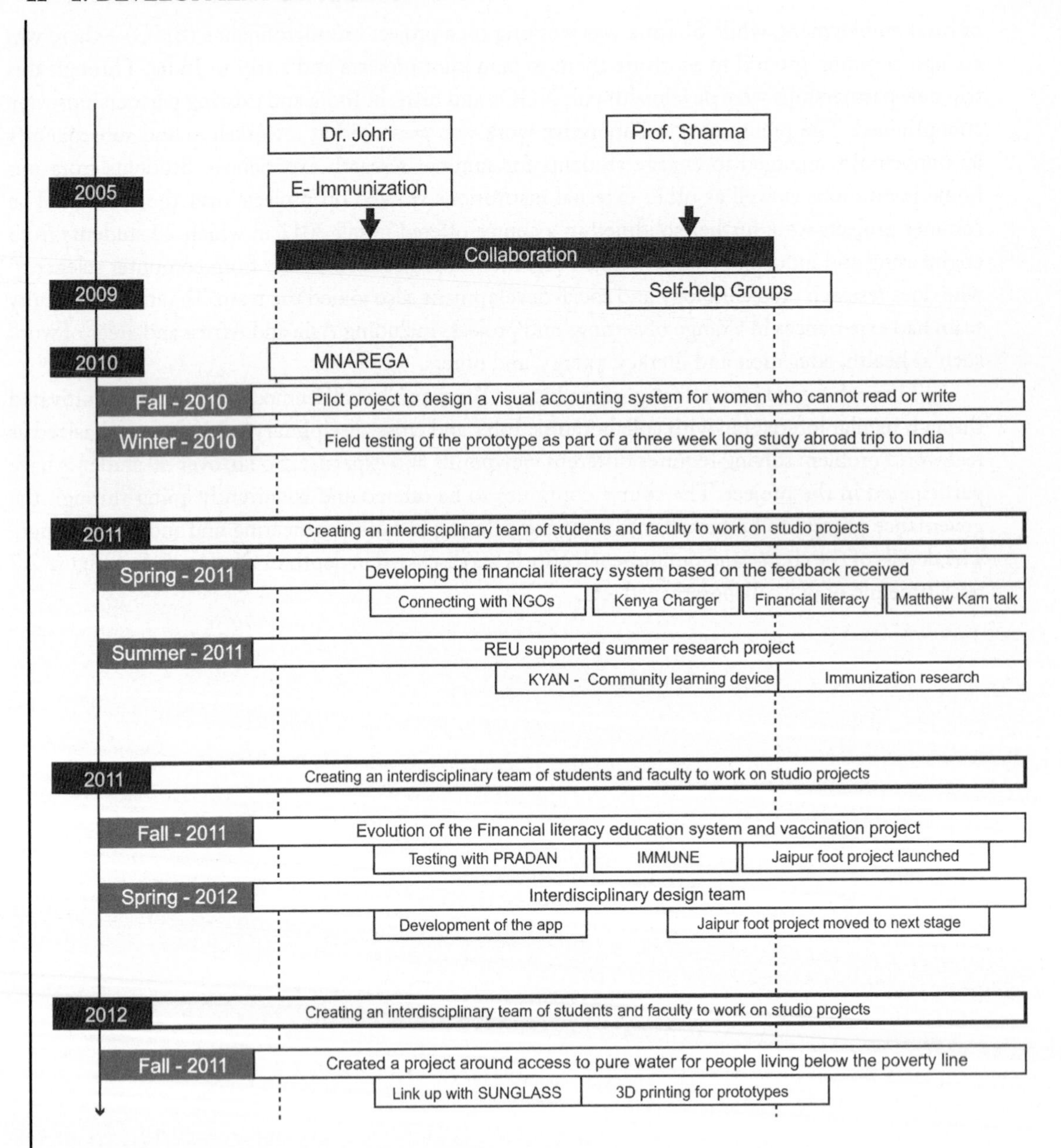

Figure 2.7: Trajectory of collaborative partnership among the co-authors and creation of an education and outreach program.

CHAPTER 3

Intellectual Positioning of the Program: Sociomaterial Infrastructures and Capable and Convivial Design

3.1 MOTIVATIONAL THEORIES

In addition to our personal motivations that provided the impetus for this work, we also delved into the current literature on service-learning, a form of education and outreach effort common among educational institutions, at least in the United States. Given that service-learning has demonstrated positive outcomes on student learning and development, we believed it was important to examine why students, and faculty, might want to engage with these projects. Our reading of that work as well as associated literature alerted us to some potential reasons why students, as well as other stakeholders, might engage with service-learning projects and what faculty can do to leverage their interest and make their engagement stronger. The first explanation for working on service-learning projects is prosocial motivation, which is the desire to expend effort to benefit other people [Batson, 1987, 1998]. When prosocially motivated, students are outcome focused and see their work as a means to the end goal of benefiting others. Another type of motivation that drives students is intrinsic motivation where they see the work as an end in and of itself [Amabile, 1993]. When intrinsically motivated, the students engage with a task or project for the sake of joy or pleasure they get from that task irrespective of the outcome. These two motivations can also be understood from the perspective of a teacher. When a teacher is prosocially motivated, her efforts are based on a desire to educate students and the outcome of student learning brings her fulfillment. On the other hand, an intrinsically motivated teacher finds enjoyment in the task of educating, in activities such as having a dialogue with students or in lecturing. In theory [Grant, 2008], prosocial and intrinsic motivations differ analytically along several dimensions but in practice they often act together, particularly in conjunction with self-determination, which I discuss next.

Another theory that helps us understand student engagement, not just with service-learning projects but with all educational activities, and is particularly useful in guiding the design of learning activities is self-determination theory (SDT) [Deci and Ryan, 1985, La Guardia, 2009, Ryan and Deci, 2000]. According to SDT three basic psychological needs—autonomy, competence, and relatedness—form the essential constituents of psychological development. Autonomy refers to actions that are self-initiated and self-regulated. Competence refers to experience of mastery and challenge and is evidenced in curiosity and exploration. Relatedness refers to the feeling of belonging and being significant in the eyes of others. SDT further suggests that people are naturally inclined to explore and dedicate much of their energies toward activities, roles, and relationships that promote these basic psychological needs. "Of importance, from an SDT perspective, the social context – specifically relational partners' support of needs—informs one's self-concept, goals, and identity-related behaviors [La Guardia, 2009, p 93]."

Prosocial motivation, intrinsic motivation, and self-determination theory form the backbone of our program. We believe that our projects should provide students the opportunity to act on their prosocial motivation while also being intrinsically motivated by the task at hand. We are also aware that the same motivational aspects drive us as well and are probably the reason why our partners in NGOs are driven in their work. Therefore, we leveraged SDT's three elements as a guide for the design of our program. We have found that design projects—where students work on designing something tangible—allows students to be intrinsically motivated as well as work toward something useful for others. Design-based projects also allow us to build significant autonomy as students in the course can work on the project without excessive oversight and own the project; projects allow each student to contribute and thereby display competence and build further expertise; and, given their prosocial motivation, there is an inherent relatedness among the students which is further nurtured as they work on the projects. We further engage with the relatedness aspect by building a community between current and former students as well as among the students, faculty, and other stakeholders. Finally, as suggested by prior research, dialogue and discussion form a significant part of the course, and interaction among students and faculty is encouraged. Brainstorming and design critiques are formally integrated into the course to create a culture of discussion and dialogue.

3.2 SOCIOMATERIAL INFRASTRUCTURES

In addition to the literature on service-learning and motivation, our work is theoretically motivated by the sociomaterial perspective. The sociomaterial perspective which consists of a constellation of theories such as actor-network theory [Latour, 2005], mangle of practice [Pickering, 1995], activity theory [Engestrom, 1999], and distributed cognition [Pea, 1993] builds on the socio-technical[1] approach. Socio-technical refers to the mutually constitutive nature of the technical and social aspects in system development and deployment. As Law and Bijker [1992] states,

> "Technology is never purely technological: it is also social. The social is never purely social: it is technological. This is something easy to say but difficult to work with. So much of our language and so many of our practices reflect a determined, culturally ingrained propensity to treat the two as if they were quite separate from one another." (pp. 305-306)

[1] `http://en.wikipedia.org/wiki/Sociotechnical_system`

Socio-technical design has a 50-year history [Mumford, 2000a,b, 2006] and its objective has been "the joint optimization of the social and technological systems" (2000b, p. 34). The socio-technical design movement gained enormous popularity in the 1970s owing to a greater awareness of the social impacts of technology and the attempt to improve the quality of life and manager-labor relationships through improved social relations by relegating technology to the background. Our program development follows a socio-technical approach in the sense best encapsulated in this description by Erickson [2009],

> "[S]ocio-technical design is not just about designing things, it is about designing things that participate in complex systems that have both social and technical aspects. Furthermore, these systems and the activities they support are distributed across time and space. One consequence of this is that the systems that are the sites for which we are designing are in constant flux. And even if we were to ignore the flux, the distributed nature of the systems means that they surface in different contexts, and are used by different people for different (and sometimes conflicting) purposes." (p. 335)

The sociomaterial lens, a recent extension of the socio-technical approach, dissolves the taken-for-granted boundaries between the social and the material and views them instead as a composite and shifting assemblage or a constellation of material and social [Orlikowski and Scott, 2008]. An assemblage is a heterogeneous configuration that preserves some concept of the structural [Marcus and Saka, 2006] but differs from structures in that it consists of elements that function well together but may be quite unlike each other and coevolve with time. An assemblage is only momentarily specific. This assemblage or configuration neither presumes independent nor interdependent entities [Barad, 2003]. Instead, all entities (whether social or technological, human or material) are inseparable. In other words, in this approach neither the social aspect of our work nor the material aspect is privileged. Latour (discussed in Orlikowski and Scott [2008, p. 455]), paints a vibrant picture to capture the essence of sociomateriality. He asks us to imagine a battlefield with soldiers all lined up. He argues that there are no soldiers without their uniforms and arms, nor are the materials of any use without the soldiers; they co-constitute each other. Although not as dramatic as soldiers on the battlefield, a similar picture of sociomateriality can be painted of our approach. There is no "development" per se without some form of materiality involved. Therefore, at one level development can be approached as a process of working with and thinking through materials. Even before recent innovations in materials for building, for instance, humans have made use of materials available to them for the betterment of their lives – whether that was building houses or boats. Even before the advent and exponential increase in the use of information technology, books have been a part of learning for over three centuries. Yet, the nature of material they used transformed their social landscape—from a small group of hunter-gatherers to a huge social structure living in a city. The materials—their availability or use—though shape the social structure, and people could be in slums or suburbia with very different social dynamics and quality of life. In essence, materiality and sociality are intertwined and shape life in different ways. This idea has several implications for design. One important issue to understand is that although in design and scholarly practices we make an analytical distinction between the material and the social, "Any distinction of humans and technologies is analytical only, and done with the recognition that these entities necessarily entail each other in practice [Orlikowski and Scott, 2008, p. 456]." Given that technology and development are closely intertwined, any effort to design or study them must encompass both. For instance, the same object such as a computing device can be both an affordance or an obstruction to accomplishing a task. The computer can aid in getting information needed to teach something to a student but can also be used as a gaming device thereby interrupting a teaching moment. The gets it meaning from practice and this meaning changes as the activity unfolds.

One implication of a sociomaterial perspective has been our focus on infrastructure and of understanding and examining it as both social and material [Bowker and Star, 1999, Star, 1999]. For us, following Bowker and Star [1999] and Star [1999], infrastructures are tools that undergird shared, learned practices. Practices here mean shared ways of being—of acting, doing, and talking – which are learned and enacted socially and are specific to a context and community. This idea is translated in our design practice as an approach whereby we focus not just on the artifact but the context of its use and the context of its users and the place of the artifact in the larger infrastructure of practices within the community. We also conceive of our infrastructure as "sociomaterial" because we have constructed relationships with people through and about materials; our teaching is about student-faculty relationship within material spaces and about using materials; the final output is a socially valuable material that relies as much on its materiality for adoption as its fit with the social and material context, and so on. Overall, a sociomaterial infrastructures conceptualization allows attention to all aspects of the project—not just the technology or the people.

3.3 CAPABLE AND CONVIVIAL DESIGN

The design of new technologies for use in low-resource environments, often aimed at fostering social or economic development, has been growing as an area of interest among researchers and practitioners in both the development realm and information and communication technologies (ICTs) communities in the past decade [Bada and Madon, 2006, Heeks, 2008, Silva and Westrup, 2009]. In particular, the exponential growth and adoption of ICTs such as mobile phones have accelerated these efforts giving rise to academic communities such as "ICT for Development" (ICTD). Within engineering, two arguments have been foundational in framing the ICTD agenda. The first is that technologies originally designed for the contexts that are infrastructure abundant are not ideally suited for developing regions and this in itself offers rich possibilities for new research and application in low-resource settings. The second argument is that the conditions of underdevelopment usually represent a need for technological intervention for new efficiencies and this in turn lays out a research and application agenda that would typically find no market takers in the industrialized world—be it low-cost telephony, new speech-based technologies in less represented languages, or interfaces for non-literate users [Brewer and Demmer, 2005]. This argument, that the needs are unique and require unique design and implementation is a foundational motivation for our work and approach.

Most design-oriented scholars argue, the goal to develop ICTs that address the needs of end users in a contextually relevant manner implies that one of the imperatives is to understand social, economic, and physical context where the product will be used. In other words, an effective approach necessitates a contextual inquiry as a designer has to work in scenarios where the past experiences with any form of technology can be limited and a larger ecology of technology is absent. User-centered design approaches lay out in detail how the actual process of design should proceed including techniques such as brainstorming and the use of prototypes and are especially applicable to contexts that are unfamiliar to designers. Scholars and designers have advanced numerous user-centered design approaches with unique. For instance, participatory design [Muller and Kuhn, 1993] and end-user development [Lieberman et al., 2006] emphasize the inclusion of the user in the design process and the tailorability of the design artifact. Contextual design [Beyer and Holtzblatt, 1998], a popular approach for user interface and system design, outlines how to account for context through a step-wise process. Scenario-based design [Carroll, 2000] is a specific mechanism that outlines how user scenarios of use are critical for useful and usable design. Even though these approaches are motivated primarily, though not exclusively, by non-ICTD contexts, each of these perspectives has something useful to offer technology design for development and can be appropriated for ICTD work [Kam et al., 2007, Ramachandran et al., 2007].

Recently, design scholars have made a persuasive argument that the values held by designers and the values reflected in the designed product [Friedman and Kahn, 2003, Friedman, 1996, Johri and Nair, 2011] should be given greater consideration in design of artifacts. To capture the importance of values and provide guidance for the study and design of technology that pays attention to values several frameworks have recently been proposed such as value-sensitive design [Friedman and Kahn, 2003, Friedman, 1996], value-centred design [Cockton, 2005, 2004a,b], and reflective design [Sengers et al., 2005]. This emphasis on values has considerable implications for design for development, specifically the Value Sensitive Design (VSD) framework. VSD provides extensive guidance on how to conceptualize the nature of values and their impact on stakeholders, how to carry out empirical investigations that can reveal the contexts of designers and users, and how to conduct targeted examination of the technological artifact in use [Friedman et al., 2006, Friedman, 1996]. The VSD framework builds on a range of prior work in human-computer interaction, social informatics, computer supported cooperative work, computer ethics, and participatory design. Its proponents define it as: "VSD is a theoretically grounded approach to the design of technology that accounts for human values in a principled and comprehensive manner throughout the design process" [Friedman et al., 2006, p. 349]. Within VSD, value refers to "what a person or a group of people consider important in life" [Friedman et al., 2006, p. 349], a definition similar to the one put forward by Sen [1999] in the context of human development. Therefore, examination of values within the VSD framework requires not only an understanding of the context but also an interpretive account of the interests and desires of people. Values are motivated not just by the situational circumstances, the here and now, in which designers or users find themselves, but also by their own and their artifacts' larger cultural-historical trajectory. Theoretically, VSD is an interactional approach where "values are viewed neither as inscribed into technology (an endogenous theory), nor as simply transmitted by social forces (an exogenous theory). Rather, the interactional position holds that although the features or properties that people design into technologies more readily support certain values and hinder others, the technology's actual use depends on the goals of the people interacting with it" [Friedman et al., 2006, p. 362]. Although useful, VSD suffers from certain problems primarily being a prescriptive list of important values: Human Welfare, Owenership and Property, Privacy, Freedom from Bias, Universal Usability, Trust, Autonomy, Informed Consent, Accountability, Courtesy, Identity, Calmness, and Environmental Sustainability (2006; p. 364-365). In practice, although design professionals definitely consider personal, social, and economic values, there may be a lack of consideration of moral values. This shortcoming of VSD has also been discussed by Le Dantec et al. [2009] in their paper about the role of context in value salience—what values are important when—and the overall benefits and pitfalls of using any form of classification. They raise the question whether classifications limit more than support value-based design and argue that VSD should provide leeway to encapsulate locally expressed values during the design process. Johri and Nair [2011] show that values often emerge in context as designers grapple with the problem at hand and have to design solutions using resources available within the design context.

In spite of their relevance and application in ICTD, the approaches identified above lack a strong basis in human development and empowerment, issues that are critical to development. Recently, Johri and Pal [2012] have advanced a design framework that comes closer to capturing the approach we, the authors, adopt in our work—capable and convivial design (CCD). From Sen [1999], (2) *Creativity/Intensification* is a dimension that focuses on the actual use of the artifact beyond just access to it for means that give the user joy, (3) *Accomplishment of Self/Others*, under which we address issues of self-respect, gender relations, power dynamics, relationships, and, (4) *Participation/Collectivity*, which brings to fore issues of a participative culture and working in a collective milieu. We use these four issues as a pathway into interpreting Ivan Illich's work in order to create a framework. Illich's idea of conviviality strongly complements Sen's notion of capabilities and provides a way to reconcile the individual-oriented nature of Sen's work with institutional structures. This allows for the creation of a meaningful design framework that targets an environment where the social context often creates a significant collective, and in turn, personal motivation [Konkka, 2003]. Combining Sen's principles of capabilities—respecting the values users think are important to them—and conviviality—giving users more power—they have advanced four primary characteristics that design should target in accordance with the Capable and Convivial Design (CCD) framework:

1. Access to artifacts (Ease of Accessibility)

2. Ability for self-expression (Expressive Creativity)

   (a) Ability to use personal energy creatively

   (b) Ability to personalize the environment

3. Ability to interact and form relationships with other people (Relational Interactivity)

4. Opportunity to enrich the environment (Ecological Reciprocity)

Johri and Pal [2012] further explain these four characteristics thus: (1) access to artifacts means artifacts are universally accessible and that the ability of one person to be able to use the tool should not take away the opportunity from another person. In a manner, the CCD framework argues for equitable distribution of resources; (2) Self-expression captures the idea that people should have the freedom to express themselves such that they are able to use their effort in a creative manner and also be able to modify the environment or the tool in a manner that is personally useful and satisfying (this element is present in many of the new web-based computing artifacts that are available today as we discuss below); (3) the relational component of the framework stems from the focus of capability and conviviality on the social aspects of human life and the ability and need for people to form ties with other people (people should be able to develop and maintain associations with others to share ideas and this augments their creativity, this is the backbone to producing and sustaining a society that values individual freedoms, and (4) the ecology component highlights the need for people to give back to their environment and not just take resources from it. In a way, people form a relationship with the environment they live in, in such a way that each privileges and benefits the other. Given the current focus and awareness of environmental issues, this characteristic becomes particularly relevant as we design new artifacts.

The CCD framework through its emphasis on the four characteristics discussed above complements and extends current design traditions discussed earlier, such as contextual and participatory design, and makes tangible ideas such as empowered design [Marsden, 2008] and stakeholder engagement [Ramirez, 2007]. Furthermore, as Johri and Pal [2012] argue, the CCD approach goes beyond other design-based approaches by making user expressivity central to development efforts and by highlighting that contextual and institutional considerations are central to development efforts. Another important characteristic of the CCD framework is that it is on the nature of empiricism and does not prescribe any specific methodology. Furthermore, it allows leveraging existing perspectives such as "design thinking" and "design-based research" and allows designers to work within the traditions of other approaches such as user-centered design and participatory design. It overcomes a central critique of VSD that it is over prescriptive and advances broad ideas that can guide design—whatever the exact design method adopted. Finally, the CCD framework brings both the design of the artifact as well as its implementation and its consequences for the redesign of the artifact with in the purview of the overall design process. Johri and Pal [2012] outline several guiding principles for developing their framework and we believe it is important to briefly review them as they outline why the framework is important and useful. These principles and the ideas of "capabilities" and "conviviality," that form the backbone of the CCD framework, are also crucial motivating ideas for our work. In advancing CCD, Johri and Pal [2012] argue that given the extremely relative nature of social and economic development in different geographies, arrival at a single set of design principles is challenging, and a useful framework must apply across contextual variations. Therefore, CCD encapsulates broad principles that are useful and effective but at the same time respects design diversity. Second, CCD serves both pragmatic and inspirational purposes—it helps guide the design process (how to design) but also provides an overarching framework to establish need for design (what and why of design). This complimentarity is critical for user empowering design as it is essential not to lose focus of the overarching goal while designing as well as not be focused on design details. Finally, the framework addresses one of the most severe shortcomings of design of technology for development—it guides new technology development that goes beyond basic access and needs and provides a more comprehensive approach that also supports imagination. A core concern that has emerged over the last few years is a disproportionate emphasis in current design for development on "basic needs" of users in low-resource environments without adequate attention to user motivated concerns that would enrich their lives from their perspective, rather than merely provide access and satisfy basic needs. In particular, scholars argue, there is a need to design products that "empower users" [Liang, 2010, Norris, 2001] to make judicious use of resources that already exist [Gurstein, 2003, Warschauer, 2002]. This idea of empowerment, which has been a critical motivation for us in our work, is supported strongly by the work of Sen [1999] and Illich [1973], the scholars whose ideas have been translated through the CCD framework. Sen argues that the primary end and principal means for achieving human development is individual freedom and Illich's idea of conviviality broadly refers to power and control of individuals over the range of physical and metaphorical tools they possess as part of their social and economic being and use of those tools for self-expression. In essence, they both emphasize user empowerment and personal freedom as the means to advance human development thereby providing an avenue for moving beyond needs and access and toward the enrichment of users' lives. We refer readers to Johri and Pal [2012] for an in-depth discussion of Sen and Illich.

3.4 LINKING SOCIOMATERIALITY AND CAPABLE AND CONVIVIAL DESIGN

There is a direct link between the position of sociomaterial infrastructures and capable and convivial design approach. As Johri and Pal [2012] explain, capable and convivial tools allow symbiotic engagement with self, others, and the environment. Consequently, these tools enhance a person's self-image as they use a tool and make it possible for them to invest the world with their meaning. Similarly, the sociomaterial infrastructures idea emphasizes the link between the social and the material and importance of addressing both the social and the material in tandem. The distributed cognition [Pea, 1993] approach in particular is well suited for understanding both sociomateriality and capable and convivial design (CCD). Scholars in the distributed cognition tradition argued successfully that cognition is distributed and is accomplished or performed through a combination of artifacts and people working together and is embedded and cannot be isolated from practice [Pea, 1993, 1994]. Therefore, one of the central actions of life—meaning-making—involves the use of material in everyday practices and is central to our ability to act on our freedoms, as recommended by the CCD framework. Consequently, any successful design project depends to a large extent on understanding the practices in which any artifact is to be used and in examining how the artifact will allow the user to shape her self-image and improve meaning-making.

CHAPTER 4

Case Study–Quick Response (QR) Code Based Immunization Solution

Up to 60% of infant mortality in developing countries can be prevented through appropriate childhood vaccinations[1] and India is no exception to infant mortality due to lack of immunization. Several converging factors contribute to a low level of healthcare in general and infant mortality specifically in India.[2] First, a large percentage of the parents are simply unaware of the importance of childhood immunizations and in surveys 50% of parents respond that they do not find it important to get their children vaccinated. Furthermore, in addition to the lack of awareness, access is also an issue and only 25% of the Indian population has access to Western medicine, which is typically practiced more in urban areas. Without access to Western medicine in rural areas and only a third of India's hospitals and health centers located in rural areas, three quarters of the population that resides in rural India is at a disadvantage. Yet, as is typical in most countries, healthcare is one of India's largest sectors in terms of revenue and employment and with India poised to become the world's most populous country by 2025, there is no indication that the sector will become less important in the future. Within this context, with India and China accounting for nearly a third of the world's child deaths, a majority because of lack of access to and awareness of immunization, any productive steps toward improving immunization efficiency can go a long way in saving the lives of many children in India. Immunization is recognized as one of the most cost-effective and highest-impact health interventions available to us. Three million people die every year of diseases that are preventable by readily available vaccines.[3] The vast majority of these preventable deaths take place in developing countries. There are many reasons that for the existence of an "immunization gap".[4] They include inadequate geographical coverage of the immunization program itself, a lack of awareness in the population,

[1] `http://en.wikipedia.org/wiki/Infant_mortality`
`http://www.who.int/maternal_child_adolescent/documents/pdfs/lancet_child_survival_reduce_mortality.pdf`
`http://www.unicef.org/immunization/files/SOWVI_full_report_english_LR1.pdf`

[2] Immunization Profile—India: `http://apps.who.int/vaccines/globalsummary/immunization/countryprofileresult.cfm?C=ind` Immunization Coverage In India, A Report by the Institute for Economic Growth, India: `http://www.iegindia.org/workpap/wp283.pdf`

[3] WHO report: State of the World's vaccines and Immunizations: `http://www.who.int/mediacentre`

[4] Defined as the percentage of children that should be immunized but are not. The immunization gap in a country like India is about 40%. In the state of Andhra Pradesh where the e-Immunization prototype was tested it is 28%.

and various inefficiencies in the program. It is these inefficiencies which are most pronounced in the areas where we can least afford them (i.e., rural and remote areas in developing countries), that need to be addressed most urgently.

Efficient immunization is a complicated undertaking due to several market and supply chain dynamics. The worldwide vaccine market, particularly for childhood immunizations, is in a state of flux. Vaccinations typically have extremely low profit margins and consequently pharmaceutical firms are reluctant to produce more vaccines than are required. The demand for vaccines in developed nations is easy to predict and given the elimination of many diseases the requirements are also limited in terms of the different kinds of vaccinations required. In developing countries such as India, this situation is far more complicated. Furthermore, vaccines rely on an extensive and complicated supply chain mechanism for delivery. They have to be stored at a specific temperature and have short expiration dates. Therefore, data about use and time frames are critical. Significant wastage currently occurs because of these problems. Another barrier to improving the delivery of vaccinations in developing countries is a lack of concrete records. When children are brought in for vaccinations without proper documentation, they are simply given another one, which is inefficient in cost and time.

Table 4.1: Schedule of Vaccinations

Vaccine	Ages of administration
BCG	Birth
OPV0-3	Birth, 6, 10 & 14 weeks
DTP1-3	6, 10 & 14 weeks
HBV1-3	6,10 & 14 weeks
Measles	9 months
Vitamin A	9 months – 3 years (1st dose is given with measles vaccination. 2nd dose is given with DPT booster vaccination)

Finally, a key problem with immunizations, related to data and information, is that some immunizations are easy to deliver as they only need to be given once before a child reaches a specific age, while others need to be given at specific time intervals necessitating more complicated information capture and use (see Table 4.1). In this situation, if the parents are not involved in the health of the child, which they often are not in rural areas, the entire imperative falls on the health workers in these areas to ensure that the child is vaccinated. In a highly populated country such as India, where each health worker is responsible for multiple villages and health records are not well maintained, it is easy for a child to slip through the cracks. This situation represents an irony of modernization in India where the technology is available, the human capital is also abundant, but the training that can sync technology and people is absent. This synergy is particularly missing for healthcare workers who are the people on the ground and can ensure that they arrive at the vaccination site with enough but not too many vaccines. They can also ensure that the children who are supposed to get vaccinated are aware and the parents realize that they were supposed to bring the health card to get proper shots after they arrive at the health center. The number of patients that need to be vaccinated is very high and the resources allocated to do the job are not enough. As a result, the primary focus stays on vaccinating as many children as possible. The result is an overworked health care worker, who is always catching up.

4.1 THE HEALTH AND IMMUNIZATION LANDSCAPE IN INDIA

Immunization[5] is carried out in most countries through programs that are run by the concerned government body as part of the existing health machinery and India is no different. The responsibility of the government of India in terms of health care is divided between the Center (similar to the federal government in the US) and states and union territories (similar to the District of Columbia in the US). Constitutionally, health is a state subject in India (Article 47 of the Indian constitution) and therefore the states have the primary responsibility for improving public health. In the urban areas health services are provided by local bodies like the Municipal Corporations, as well as by private practitioners / institutions. The rural health services are managed through a network of community and primary health centers and sub-centers, each catering to a defined population and area. The government of India is responsible though for the planning of health services and programs and for providing funding as well as technical and material support. As a matter of fact, the bulk of the expenditure on public health programs is borne by the government of India, through its own resources, with some external funding. Importantly, all preventive and health promotional services provided to the general population through government hospitals are free. All immunization services are provided under the umbrella of the Family Welfare Program (FWP). This program was initiated in 1952 and at that time it was primarily a clinic-based family planning program. Subsequently, it was integrated with the Maternal and Child Health (MCH) Program and its activities broadened to include a variety of services to mothers and children, including antenatal, delivery, and post-natal care, immunization, and counseling on maternal and child health and nutrition. Specific emphasis was given to reducing maternal and child mortality through a focus on treatment of childhood diarrhea, neonatal care, and strengthening of emergency obstetric care.

In India child and maternal healthcare is closely tied to the issue of population, and the National Population Policy released in February 2000 emphasized the need to strengthen health care and education of women and children to achieve population stabilization and improve the quality of life of its citizens. The policy document also clearly states that one of its key strategic goals as part of the Action Plan for Child Health and Survival is to "ensure 100 % routine immunization for all vaccine preventable diseases, in particular tetanus and measles."

[5]The data and primary issues covered in this section come primarily from two reports: "Country Proposal for Support to the Global Alliance for Vaccines and Immunization and the Vaccine Fund" and "Synthesis of Immunization Assessment in India"

Planning in India is done through five year plans, and the ninth five year plan emphasized the need to intensify efforts to increase participation of general medical practitioners in voluntary, private, and public-private partnerships, especially in the urban areas, where the public health system may have lower coverage. In India, overall, there is significant emphasis on public-private partnerships across all areas of governance and public service, and the Indian Ministry of Health works closely with professional bodies and private associations such as the Indian Academy of Pediatrics, the Indian Medical Association, at both the national and regional levels, and private sector health institutions. In the last decade, it has initiated programs in collaboration with professional associations of private industry for the involvement of industrial organizations in health messages to their workforce, such as family planning, HIV/AIDS awareness, and maternal and child health camps.

Table 4.2 lists vaccination targets for recent years in India. Table 4.3 presents data on wastage due to supply chain issue and because of dropout by targeted individuals due to issues of migration or data deficiency. Finally, Table 4.4 lists the expenditure by the government of India on immunization programs.

Table 4.2: Vaccination Data and Targets for India

Number of (In Millions)	Baseline	Targets			
	1999	2000	2001	2002	2003
Births	26.30	26.78	27.26	27.75	28.25
Infants' deaths	1.84	1.82	1.80	1.78	1.75
Surviving infants	24.46	24.96	25.46	25.98	26.50
Infants vaccinated with BCG*	16.63	18.22	19.86	21.56	23.32
Infants vaccinated with OPV3**	14.92	16.47	18.08	19.74	21.47
Infants vaccinated with DTP3**	13.45	14.98	16.55	18.18	19.88
Infants vaccinated with Measles**	12.23	13.73	15.28	16.89	18.55
* Target children out of total births ** Target children out of surviving infants					

In Table 4.2, the baseline is calculated using evaluated coverage for 1999, based on the *WHO/UNICEF Review of National Immunization Coverage 1980-2000 India, November 2001* estimates. For Tetanus Toxoid coverage, the baseline is from coverage estimates made by the National Family Health Survey II for 1998-99.[6]

[6] Formula to calculate DTP vaccine wastage rate (in percentage): [(A—B) / A] x 100. Whereby : A = The number of DTP doses distributed for use according to the supply records with correction for stock balance at start and end of the supply period; B = the number of DTP vaccinations.

Table 4.3: Wastage and Dropout Rates of Vaccinations in India

	Actual	Targets		
	2000	2001	2002	2003
Wastage rate [36]	25	25	20	20
Dropout rate* [(BCG - DTP3) / BCG] x 100	12	10	10	8
Hepatitis-B Vaccine Wastage rate	-	-	20	20

** Drop-out rate has been calculated using BCG-DPT3, due to non-availability of evaluated coverage data for DPT1.*

Table 4.4: Estimated Recurring Expenditure by Government of India (GOI) on Immunization

Estimated GOI Recurring Expenditure on EPI: 2000-01 to 2002-03 Rs. Million				
Major Component	*2000-01*	*2001-02*	*2002-03*	***Total***
Vaccines	1173.6	1349.6	1552.1	**4075.3**
Salaries	760.8	874.9	1006.2	**2641.9**
Maintenance incl. cold chain	369.6	425.0	488.8	**1283.4**
Supplies & Construction	78.0	89.7	103.2	**270.9**
Transportation	72.0	82.8	95.2	**250.0**
IEC	30.0	34.5	39.7	**104.2**
TOTAL	**2484.0**	**2856.6**	**3285.1**	**8625.7**
USD Million	(52.9)	(60.8)	(69.9)	(183.5)

4.2 PROBLEM FACED BY IMMUNIZATION PROGRAMS IN INDIA AND AN ICT-BASED SOLUTION

The background for our effort into developing an immunization solution had its origins in a previous project—e-Immunization[7] (also referred to in Johri's personal trajectory).

[7]The e-Immunization team was led by an officer of the Indian Administrative Service, who has overseen many implementations of technology initiatives for government programs, and included seven graduate students from Stanford University including one MBA student, three engineering students, and two students from the School of Education.

Rajendra Nimje: An officer in the Indian Administrative Service, which is the backbone of Indian Administration, who served the Govt. of Andhra Pradesh in various capacities. He was serving as a Reuters Digital Vision Fellow at Stanford while leading the 'e-Immunization' project.

Anshuman Bapna: An MBA student at Stanford Graduate School of Business with a B. Tech in Electrical Engineering from IIT Bombay. He founded and was CEO of a 21 person technology venture in India that worked on online collaboration technologies.

As part of the e-Immunization project several problems with immunization efforts in rural India were identified. The project also identified that to improve immunization any innovation had to target the health workers. Specifically, the innovation would have to target the following issues:

1. *Excessive paperwork*. Health workers report that the two activities on which they spend the most of their time are report writing and ledger filling. Both activities take time away from what they need, and want, to do to be more effective. (For example, it can take a health worker up to *15 minutes* to record an immunization that took *2 to 3 minutes* to administer.)

2. *Paucity of information*. A health worker in a rural area spends most of his/her time in the field and is away from any source of information for a significant portion of the time. Another issue we uncovered was that incorrect or outdated patient information was available to the health workers, which prevented them from being more proactive in their administering of the program, as well as made mistakes more likely. Furthermore, the rural population migrates continuously which can compound this problem.

3. *Tremendous workload.* Immunization is just one of the health workers' many responsibilities.

Our research also uncovered that many parents were misinformed or unaware of immunization, as several studies had reported. These parents did not come back for the next dose of immunization and dropped out of the system. In the area we studied, of children aged 12-23 months, 42% are fully vaccinated, 44% are partially, and 14% never. These partially vaccinated children become a priority target in our first generation solution. Finally, in addition to health workers and parents, we also found that there was inadequate monitoring of performance at the individual and district levels. This was compounded by a lack of accountability and transparency at different levels in the system and there was improper spending or focusing of funds by the responsible agency due to a lack of available data. In order to effectively monitor and measure the success of immunization programs a large amount of data needs to be maintained, and this is currently not done.[8] Finally, loss or wastage of resources (mostly vaccines lost due to theft, improper maintenance of inventory because of insufficient data, and an inability to maintain the cold chain[9]).

Amita Chudgar: Doctoral student in education at Stanford. She has a masters in Economics from Mumbai, India, and a masters in Development Studies from Cambridge, UK.
Aditya Johri: PhD student at School of Education, Stanford. He holds masters degrees in Information, Design, and Technology from Georgia Tech. and Mass Communication from the University of Georgia. He also holds Bachelor of Engineering from the University of Delhi. His doctoral work looked at educational uses of technology.
Jagannath Krishnan: Has extensive experience in software engineering, which includes work at Tata Infotech, VMWare, and Lucent Technologies. He has degrees in Computer Science from Goa University and Stanford University. He also has extensive experience working with multiple charities to further child education in India.
Satyajeet Salgar: Software engineer working in the Bay Area. He has a B.E. in Computer Engineering from the University of Pune and a masters in Computer Science from Stanford University. His engineering experience includes work in a medical startup spun out of Stanford and a security startup.
Srihari Yamanoor: A mechanical engineer pursuing a PhD at Stanford University originally from Andhra Pradesh with great interest in health issues in India.

[8]In the state of Andhra Pradesh, for instance, the survey to identify pregnant women is carried out just once a year in April.
[9]The cold chain refers to the need to keep vaccines consistently at a temperature of 2-8 degrees centigrade.

4.3 THE E-IMMUNIZATION SOLUTION—ICT BASED IMMUNIZATION SERVICES

The e-Immunization project leveraged information and communication technologies to address these shortcomings in currently operating immunization programs. The e-Immunization solution was built around a centralized database that contains the immunization history and schedule for each child. Health workers used specialized handheld devices that update this database regularly from the field. The device also prints out immunization receipts for parents and updates a **smart card**, which is given to each child and contains his/her immunization history. The adoption of this system by existing immunization programs will lead to a significant improvement in immunization coverage. The e-Immunization project focused on increasing immunization coverage by targeting increasing health worker productivity, allowing the focused targeting of each child, and ensuring parents are better informed during each stage of the immunization program and allowing effective monitoring at every level of the immunization program hierarchy. The e-Immunization project completed a prototype implementation in Andhra Pradesh, India.

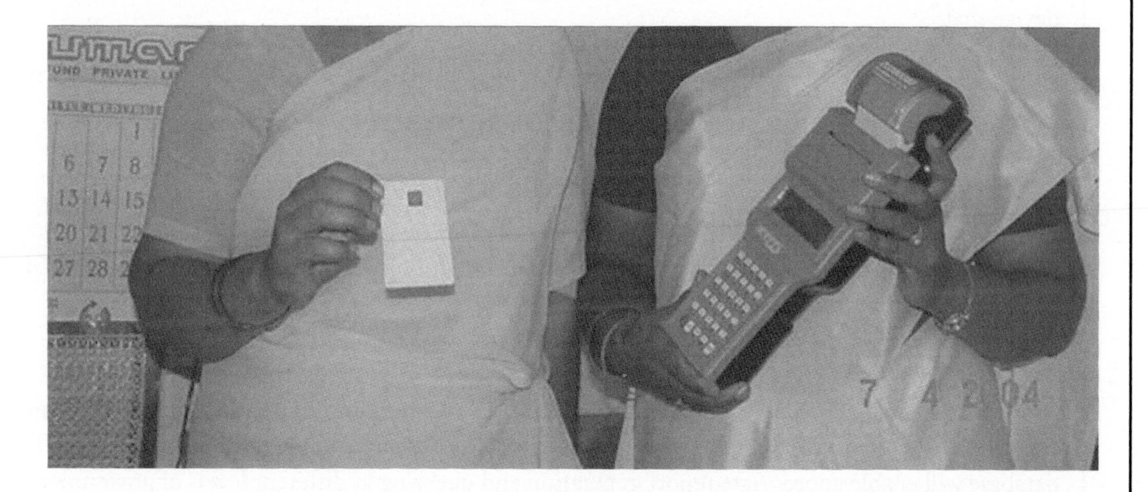

Figure 4.1: Health workers with Smart Card and handheld device during field testing of e-Immunization.

4.3.1 PROTOTYPE TESTING—SETTING AND IMPLEMENTATION

The mechanism for administration of vaccines in immunization programs tends to be hierarchical. For example, in the state of Andhra Pradesh (AP) in India there are 23 districts, with an average of 60 Primary Health Centers (PHCs). Each PHC has an average of five MPHAs (Mobile Primary Health Assistants) or MPHWs (Mobile Primary Health Workers). The leaf at the end of the tree is generally a single health worker responsible for a group of villages (occasionally it may be a team of two). Immunization is just one of this person's responsibilities.

Out of 20 days each month, MPHWs spend 13 days in the field, four days at Sub-Center, 4 days at PHC, one day at district headquarters.[10] At these times, they have a specific objective (say, administering to a diarrhea outbreak) and are unable to record other health details they come across. With a handheld device, the probability that they would be able to record data increased.

Data mining could then generate alerts for preventing outbreak of diseases. Furthermore, the lure of "possessing" a Smart Card would be a useful marketing tool to attract greater populations to visit the Sub-Center. More importantly, the smart card allows the user government to tie in other highly prized benefits in the future—purchase discounts from government-owned grain and fertilizer stores depending on "immunization score" on a card is one of the intriguing ideas.

Overview of Technical Solution The solution made use of information and communication technologies to address all issues identified in the field research. A centralized database contains the immunization history and schedule for every child and will be updated directly from the field by the health machinery of the implementing government. Each health worker is equipped with a handheld device that:

1. stores immunization information relevant to the health worker's area
2. prints out receipts for parents, for each immunization visit
3. reads from and writes to a "smart" card that will serve as the immunization record for each child and hold each baby's immunization history
4. updates a server with all the immunization information at the end of the immunization day, directly from the field
5. can query the server for patient information.

Parents will be issued "smart" immunization cards containing immunization information of their child. The health worker will upload data directly from the field to the centralized database. This database will enable appropriate report generation and querying at different levels of government to track immunization coverage, ensuring accountability of the health machinery.

A rural district in the state of Andhra Pradesh was chosen as the pilot test site to test the e-Immunization solution. This district was also the subject of a World Bank funded project called infoDev started in 2000 that aimed to introduce IT to the healthcare services in this area. While this project failed to take off for various reasons, it sensitized the government machinery and healthcare workers to the potential role of information technology in healthcare. Health workers in that area were used to using a PDA based device in their work previously and therefore had some relevant training prior to participating in our pilot. Overall, the results of the pilot testing were very encouraging. Our handheld device which we named the "immunizer" was much faster in retrieving

[10] From "Role and Efficacy of MPHW & Male Health Worker in Andhra Pradesh," a study commissioned by DFID-India, 2003.

records—10 to 30 seconds against the PDA which used to take 2 to 3 minutes. Reducing this time lag made a significant contribution to the success of this project as we found that parents used to put pressure on health machinery for quick immunization. Health workers used to yield to this pressure by inoculating the baby first, writing down the information on a piece of paper, and then thinking of putting the data into PDA which of course meant double work and hence was often completely avoided. The health works also received Smart Cards with great enthusiasm. The Smart Card does away with the need to search the records for a child or to manually enter records for a child on a card (for the family). This whole process of manual search and entries of records used to take almost up to five minutes against the inoculation time of 20 to 30 seconds. The attached printer to the "immunizer" has also been very well received by the health workers and parents. It prints the complete immunization history of a child in 10 seconds which is handed over to the parents. The health workers also informed us that they found the keyboard operation of "immunizer" a lot easier compared to stylus use in PDA which they were never comfortable with.

4.4 URBAN POPULATION AND IMPROVING IMMUNIZATION WITH ICT

Almost five years subsequent to the e-Immunization project the authors revisited the idea. One central motivation was that healthcare and immunization in particular is still a critical problem in India. Furthermore, the past decade has seen a revolution of sorts in the proliferation of mobile devices in India and we were interested in examining how this changed environment might be leveraged. Given that prior work on e-Immunization looked exclusively at rural contexts, we had considerable understanding of the issues there. Therefore, we wanted to study a diverse perspective from another context that struggled with the same issues but was geographically different. This time we decided to look at urban slums that have many similar problems even though they are embedded in highly "developed" urban areas.

4.5 UNDERSTANDING AND EXPLORING THE DESIGN CONTEXT

The initial conversation between the authors led to an ethnographic study in a slum in Jaipur, Rajsthan, India, to understand the different aspects of life in an environment with poor health records. We did not start with a focus on record keeping, but wanted to just understand the living conditions, expectations, and desires of the user group that presents a severe problem in information management in general. We decided to investigate a community of about 2,500 families, almost all of them engaged in puppet making, an ancient craft based profession in India. Most of the families have migrated here from one particular region in India, and currently it is the second generation of migrants residing in the community. We used the following categories to understand life in the slums. The data was collected using interviews, still images, and notes. Another noteworthy aspect of the data collection was the premise under which it was conducted. We embedded ourselves as part of the government census team to assist in the process of collecting information about the residents of the community. To understanding the micro level details of the life in a non-regulated community, it is important to understand some general observations about India and its diversity. India is home to 25 languages and 2,500 dialects. It is not a simple task to comprehend a system that would work across languages and cultural icons. The different stages in an immunization schedule for children makes the situation more challenging. Research has shown that regulated residential development in either urban or rural areas is relatively easier to provide services that need to be delivered in stages, like vaccination. The challenge becomes nearly insurmountable when the target population is floating from one location to another and at the same time also changes in size based on various factors like employment, festivals, and seasons. If one looks at the point of view of people living in the unregulated residential developments, it is no surprise that vaccinating their children is not at the top of the list that they need to take care. First comes food and everything else follows where one gets food from. It is human nature to reap the benefits of hard work and, naturally, once income stabilizes, it is time to invest in transportation, communication, and entertainment. These are not mutually exclusive domains in the lives of people living in no regular dwellings and often there is an overlap among different categories. With this in mind, the living environment was examined using a methodological tool borrowed from anthropology that provides structured observational guidelines that can assist in exploration from the perspective of design. It also uses a simple mnemonic "AEIOU" for actions, environment, interaction, objects, and users (see Table 4.5). A critical point to remember here is that it is important to go in-depth in any one context. Although the context of the application might vary later on, important lessons can still be abstracted if in-depth investigation of a context is undertaken.

Our observations took place in Kathputli Nagar (Town of Puppeteers)[11] which is home to approximately 5,000 pappet makers, puppet shows conductors, snake charmers, and other street performance artists. The original inhabitants migrated from a region in another state in northern India and found a favorable environment in Jaipur with its influx of tourists which provided better opportunities of making a living for the traditional artists. The colony now has second generation artisans and is home to performers who have won national and international accolades. It is important to acknowledge here that the unorganized nature of buildings and the apparent lack of modern facilities do not translate into lack of opportunities or aspirations. The observation sessions were conducted in May 2010. The temperatures stay in the 100 degree range with the highs reaching 120 degrees on most days. A digital camera, notebook, and a cell phone for videos was used. Before taking pictures, verbal approval was secured and the participants had an opportunity to check the images.

[11] More about Kathputli Nagar: `http://www.saarthakindia.org/kathputli.html`
More about Kathputli of Puppets: `http://en.wikipedia.org/wiki/Kathputli_(Puppet)`

Table 4.5: Framework for Understanding User Context

Framework for Understanding User Context	
Actions	What are the actions taking place in the community?
Environment	The different components of the environment the community lives in and some of the "watering holes" where a large concentration of users take is the environment a factor in different aspects of their lifestyle?
Interaction	What interactions are unique to the community and how do they help identify some potential design opportunities?
Object	What are the objects with which users interact?
Users	What are the different user profiles?

4.5.1 ACTIONS

The general atmosphere in the community is of a laid-back nature. Women are in charge of household duties like cleaning, cooking, feeding, and keeping the household in order. They are also responsible for drinking water and have to go to the community hand pump and fetch water in 5-6 gallon plastic containers. The men are either at work, performing puppet shows at different venues or on the streets, or they are at home watching TV, sitting around chatting with friends, or just hanging out. The children are engaged in numerous activities—playing on the streets, watching TV, or helping their mothers with housework, depending on their gender and age. The younger kids run around, play, fight, and generally have fun till they get hungry and at that point they come home. The most common occupation is puppet making and performing puppet shows along with cleaning, sweeping, and rag picking. Most of the occupations are low skill and require manual labor. The actions observed during the study provide a basis for the laid-back lifestyle that is not as structured as we are used to in the developed world. That also means that there is no specific time for the men to leave for work or a time for them to come back home. This translates into cooking times that are linked to when the men are leaving for work. These actions also vary based on the time of the year, so during winter months, when the city witnesses a rise in number of tourists, more men are out of the community for work for longer hours. During summer months, the number of tourists declines, resulting in more men sitting around killing time.

Design Insight From a design point of view, it made us realize that no single aspect of their lifestyle could be considered a reliable routine. If something was important, they made sure it got done but it had to be important from their point of view.

4.5.2 ENVIRONMENT

The slum studied as part of the ethnographic study is situated in the middle of a very well developed residential and institutional area in Jaipur, Rajasthan, in India. It is about two blocks from the legislative assembly building. It is facing a major road, and to hide the "unsightly" dwelling units, the government has built a wall about 10' tall that effectively acts as a boundary for the neighborhood.

The buildings are mostly single story with each dwelling unit having a small open space either within the building envelope or adjoining the street. There is only pedestrian traffic possible because the width is about six feet at the widest points and two feet at the narrowest point. The streets are mostly unpaved with a few sections paved with leftover building material from a construction activity. They are more like an extension of the houses, with steps acting as public seating for a casual neighborhood social interaction. It is a network of streets that weave through clusters of houses and lead to social nodes like the neighborhood grocery store oi the doctor's clinic. There is an open drainage system and it creates a messy situation during a monsoon. As it is evident in the accompanying image, the neighborhood is not planned along a system of carefully laid out streets. The houses are mostly mud and/or brick walls with corrugated iron sheets for roof, kept in place with stone slabs. Most areas in the household are multipurpose in nature. A corner in the little courtyard becomes an area for taking a bath in the morning and later it is the place for the family goat to get tied for the day. There are no toilets inside homes and all residents use an open field next to the township. They also use it for trash disposal. Since the area is categorized as a slum, there are no running water connections, street lights, or trash pickup by the government agencies. Most houses have electricity connections but it is not uncommon to have 6-8 hours of power cuts per day.

Design Insight From a design insight point of view, a dwelling unit, no matter how small or how temporary in nature, represents a focal point in their life. They guard it with zeal, try to make it better at every opportunity they can get, and want to convert that into a permanent residence. The residence provides a constant in an otherwise migratory and flexible lifestyle.

4.5.3 INTERACTIONS

In this part, we look at the kind of interactions that take place in the community. The users are talking to each other, reacting to objects like cell phones and televisions. The interactions among different age groups are very complex as there is usually one person in the family who is the primary source of income. This usually results in one way communication of him with the rest of the family members, in most cases irrespective of the age difference. Community level interactions are triggered by events that need participation of a number of individuals, like filling up drinking water first thing in the morning, or getting ready to head out to a puppetry performance that a team of artists from the community is going to perform. These usually take place in the early hours of the day. It is safe to assume that most community level interactions are dictated by activities that require participation of a large number of individuals, voluntary, like getting together to perform or make puppets, or attending a meeting to put forward a united front to a local government official. Most of the regular community level interactions take place during the morning hours, before the earning members leave for work or in the evening hours, when they return.

Neighborhood level interactions take place at the social interaction nodes in the community. Some of these are essential like the drinking water tap and some are incidental like a visit by a clothes vendor or a festival in the community. Another venue is the neighborhood grocery store which not only supplies provisions like sugar, flour, spices, and cooking oil but also daily use items

Figure 4.2: The three map images show Kathputli Nagar (marked by A)—First two images show its location with the city and the third zoomed-in image shows the density of housing within the community.

like batteries, cell phone recharge coupons, stationery items for children, etc. There is a doctor's clinic in the community as well. It provides primary health services as well as education about issues related to health and development. It is the go-to place for almost all issues related to life in the slums. The doctor is also a leader and a respected member of the community. We felt he would be a good person with whom to discuss any future proposal.

Interactions among children are dictated by either food, play, or doing what the parents have asked them to do. TV plays an important role in their lives and the portable game systems like PSP have started to make an appearance. The community has a sense of collective responsibility toward children, so mischief, if caught, will result in a scolding by an elder, related to the child or not. There is a lot of running around in the streets, which is safe because there is no vehicular traffic. A lot of time is spent in watching TV and then playing games based on some of the TV shows. Once the children are old enough to have the requisite motor skills, almost every child starts a lesson in the basics of puppetry. These lessons are responsible for a large number of young children wearing a bandage over their forearm because it gets used a lot during these lessons.

Interactions among children and adults are more or less dictated by the age of the child. The really young ones are cared for by the elders as well as their parents. The toddlers are mainly trying to imitate the younger kids and the young kids are running after young adults. Puppet making or puppetry is the end goal or destiny in most cases. Another very important interaction between children and adults is at school, since almost every child attends the primary school in the community as it is free and it provides for a free mid day meal. The free meal is part of a government program that feeds 40 million children every day, to encourage higher enrollment in the primary school system.

The authors observed the positive impact on the psychology of the children as well as adults when they saw a significant value addition to their life. In the case of the mid day meal scheme, the main goal is one nutritious free meal and the side benefit is better education.

4.5.4 OBJECTS

Communal Objects

Community tap is a very important object especially when there is water supplied through it. Almost every single household has at least one member visit the tap every day. It is the only source of drinking water for the community and often becomes a source of fights and arguments among members of the community. Street furniture like benches, electrical posts as notice boards, trees, and steps outside homes provide interesting opportunities for neighborhood level interactions among women, children, and vendors of different services and products. In fact, most of the interactions with a person unknown to the family take place on the steps outside the main entrance. It could be a government official conducting a survey or a health care worker administering shots to children. The neighborhood convenience store serves an important role in the community by acting as a common point of interaction in the daily lives of people living in the community. This invariably leads to spontaneous interactions that sometimes translate into planned community events especially during the festival season. The dairy shop serves a similar function but mostly during the early hours of the day when a member from each household goes to buy milk for the day.

Neighborhood Level Objects

Steps outside homes serve an important function. Women and elders can be seen sitting outside their homes keeping an eye on the children playing in the streets as well as striking a casual conversation with a passerby. A rapid increase in the number of cell phones is making traditional phone booths extinct but they still provide a unique functionality. One can still walk up to the phone booth operator, give them a slip of paper with a phone number on it, and ask them to dial it for them. It is a direct result of high population density and the close knit nature of the community. One can always find someone who knows how to read and write and just ask them to read a note or dial a phone number on the cell phone. We feel that this reason is responsible for the continued existence of manned pay phone booths in non-regular communities like Kathputli Nagar. Television plays an important role in not only providing information but also acts as a source of entertainment. Its functionality is usually hampered by the sporadic power supply. The tobacco and betle nut shop (`http://en.wikepedia.org/wiki/betel`) acts like a watering hole for most men, as the majority of them chew tobacco. These shops also sell candy, cell phone recharge cards, and small daily use items like soap, hair oil, and refrigerated beverages. Tea stalls also serve a similar objective but they are usually focused more on freshly brewed tea served with some snack or biscuits (cookies).

Personal Objects

Cell phones are becoming increasingly common but have introduced a very unique behavior. Since incoming call are free, most users receive a lot more calls as compared to making calls. There is a very interesting phenomenon which revolves around the concept of a missed call. Radios were on their way out but have made a resurgence since manufacturers have integrated FM receivers in the cell phones. Most cellphones popular among masses come with a reasonably loud speaker and are used as an audio player. Storage containers are used as multiple objects. They become pieces of furniture, storage units, as well as stepladders when needed. Small children can be seen to use them as things to play with. Religious idols are a common sight, especially in the user group we are focusing on. They represent not only their faith but also the families' origin and the region they came from. A storage box with a lock has a very important role in the household as most of the valuables are kept in it. It's always under lock and key and is safeguarded by members of the family. Children usually have a fair number of toys but the bigger kids usually have more since they can forcibly take them from the younger children. Most of the toys are small in size, inexpensive, and an important part of the life of a kid. Portable cots and beds are common. The authors observed the omnipresence of these small toys and how much the children valued them. In a situation where almost every other aspect of life is temporary or lacks permanence, like the physical address, identification cards, etc., a toy valued by the child seemed like a slightly more permanent fixture in their lifestyle. When thinking about creating a reference point that can be linked to the vaccination records, a toy seemed like a good option. (This option was later tested as part of the design process.)

Clothes present a very interesting picture. Most women folk wear traditional dresses but the styles are inspired by latest Bollywood movies. Men on the other hand wear jeans and t-shirts or shorts. A lot of them have names of American universities. It took little fact finding to understand the reason behind this unique aspect. The clothes that are donated to charitable organizations in US and other developed societies somehow make their way to the street side shops selling used clothing. It is not uncommon to see a person wearing UCLA or MIT sweatshirt with Levis jeans in Kathputli Nagar.

Tools

Flashlights are a common sight as well and a very important part of daily life in this community. Regular power cuts lasting couple of hours every day is part of life, and a flashlight becomes a necessity at night. Cooking stoves are an integral part of life; without these, it would be impossible for a family to survive as women in the family cook three meals from scratch every day for all of the family members. Some other objects that are important in the day-to-day life in Kathputli Nagar are buckets to store water when they get water supply in the municipality taps. Small mirrors usually hang on a wall for getting ready in the morning and a shelf or an alcove in the wall that is used to keep toiletries, hair oil, eye liner, and a comb. There is usually a bathing and clothes washing area that is in the courtyard. It is an open washing area during day time, but when an adult needs to take a bath, there is a curtain hanging on a wire stretched across the walls that can be drawn for some privacy. In this area, one can find a wooden bat to beat the clothes while washing them, a bucket, a Plastic mug, a plastic scrubbing brush, and soap to wash clothes. A sewing machine is also a common object in most households and it usually is a wedding present from bride's family. It is a heavily used object and one that is under the control of the women folk. Some of the households have room cooler (air cooler) to circulate moisture rich air in a room. They are more of a status symbol rather than utility object because of the regular power cuts that some times last couple of hours. Ceiling fans or table fans are a common sight and these are also wedding presents from the bride's family. Every household had few books in various states of degradation that are used as study material for children. Besides these, each household has shoes, slippers, and sandals for the every family member but it is not uncommon to see very young children walking around barefoot as well.

Miscellaneous Objects

A wide variety of miscellaneous objects can be found in the community. These do not fit any specific category but are omnipresent. Their existence can be linked to the no-waste philosophy. Each of the objects listed below starts out as part of one process, like construction activity, and the leftover material is repurposed as part of another process like repairing a drain or just making the yard level. The situation is changing rapidly. With the exponential increase in availability of fast moving consumer goods (FMCG) and increasing income, the reuse and repurposing of leftover objects is becoming piles of trash in the streets and alley ways. Some of the objects that can be found in the community are listed below:

1. Remnants of building material in the street
2. Leftover pieces of wood from the puppet making shops
3. Plastic bottles
4. Plastic containers
5. Plastic bags
6. Old newspapers

4.5.5 USERS

Children—General Description

Children between the ages of 4 to 9 go to the government-run primary school when school is in session. They have school bags with books and notebooks in them and most come home after lunch. In the morning before they leave for school, their mother feeds them a meal and they are provided with lunch as part of the mid day meal scheme. After school most of the time is spent playing in the streets or watching television. The children can be divided into three general categories, as these are linked to their behavior. The youngest ones are the preschool age children, i.e., younger than five years. They spend most of the time around their mothers and elders who stay at home till the primary school children come back The preschool children interact and play with the school children. There is a general atmosphere of happiness and a consensus that puppet making or doing puppet shows is what they will do once they grow up. It is especially true for boys. Young girls attend the primary school as well, but once they come back, they usually stick with their mothers and play in the vicinity of their mothers.

Girls between the ages of 7 to14 help their mothers in the daily household tasks. They start with helping with the first meal of the day around 9:30 – 10:00 AM. This meal is breakfast and lunch bundled into one. At that point the men usually leave the house and either go to work or sit in the market place which is at the edge of the neighborhood. Young girls help with cooking, washing dishes, getting groceries from the neighborhood shop, and taking care of their younger siblings. They also learn sewing so that they can make clothes for the puppets and contribute to the family income. Eventually, the majority of the residents in the community get engaged in the profession of puppet making or performing with puppets. It is a known source of income and there is no reason for them to switch to a different direction of livelihood.

Boys leave for school after eating the morning meal. Once they return from school it is either TV, playing in the streets, or following the older boys. The young boys might get asked to help with some errands, but unlike girls, boys are not as engaged in the household tasks as girls are. Young men between the ages of 14-25 have mostly dropped out of school. They have mastered the art of

Figure 4.3: A series of eight images from Kathputli Nagar depicting the products designed by its inhabitants, the unorganized infrastructure, and resourcefulness in terms of power and cooling requirements. *Continues.*

Figure 4.3: *Continued.* A series of eight images from Kathputli Nagar depicting the products designed by its inhabitants, the unorganized infrastructure, and resourcefulness in terms of power and cooling requirements. *Continues.*

puppet shows and entertain tourists at popular spots in the city. Their day does not start till noon, since most have an active social life that goes into the late hours at night. The younger boys adore them because they usually have fancier phones, some of them have portable video game devices, and they often buy candy for the younger ones. Some of them have been integrated into the puppet making business and work at home as part of the family business.

Most married men are also part of the puppet making or performance business. They have the added responsibility of children, so need to work harder. They are the ones with experience in the profession and a focus on income which is not the case with younger males in the community. The older males in the community stay at home or venture out to the neighborhood market for a cup of chai (Indian tea). These are the original settlers of the community and have seen it grow from few huts to a sprawling community of 5,000. They do not have an active role in running the day-to-day affairs of the families and in some cases are dependent on the younger members who are earning income and supporting the family. They cannot work in the puppet making business because they cannot do it fast enough; they are not able to perform in the puppet shows because it is a physically

Figure 4.3: *Continued.* A series of eight images from Kathputli Nagar depicting the products designed by its inhabitants, the unorganized infrastructure, and resourcefulness in terms of power and cooling requirements.

demanding activity. As a result, most of them spend their time at home playing with grandchildren and watching TV. Young unmarried women help their mothers sew clothes for the puppets being made in the shop and take care of the younger siblings. Married women are full time caretakers of the family. It is a very difficult existence that starts very early in the morning with the responsibility of fetching water from the community tap and ends late night with a final cleanup of the kitchen. The majority of the day is spent in either cooking a meal or preparation for it.

4.5.6 EDUCATION LEVELS IN THE COMMUNITY

The community is an excellent example of education that is driven by need. It is also a great example of distinction between education and literacy. A significant number of adults can speak fluent English, some French, German, and Spanish but most do not have basic writing skills. Their fluency in spoken English is driven by the need to communicate with tourists which impacts their earning. All kids attend the local primary school because of the free mid day meal scheme. After that the learning is driven by the professional requirements of puppet making. Girls get busy with their mothers and boys follow the older boys who are getting trained in the art of puppet making as well as performing puppet shows. There is one very interesting aspect about this community, which is the surprisingly high number of young males who can converse in English. This can be attributed to their interaction with tourists from Western countries which necessitates learning spoken English. It is usually starts with few sentences but in about a year or so they can hold a conversation in English. It is a unique community in the sense that it is homogeneous in its character. Most adults work in the same profession; they all share a similar background and cultural heritage. It is also a floating population in the sense that if a family finds a better, short term opportunity at a different location, they would usually move there for some time. Surprisingly, this situation can be seen in other communities with similar economic backgrounds but engaged in a different profession.

This is a happy, content, and close knit community that still follows the traditional model of a joint family in Indian society. The elders and women mostly stay home and take care of the family. Young men go out and work to support the family. Children grow up as a floating population with collective responsibility by the community. They play and move in the community as a group from one interesting spot to the next. The life in the community revolves around earning livelihood, making puppets, and waiting for opportunities. The population of the community changes depending on the season and during summer months it reduces but picks up again during the winter months with the arrival of tourists as well as other social activities like fairs and weddings in which the puppeteers can participate. The community is migratory and when the parents travel for livelihood opportunities, children do as well. This means that some children will be born outside the village, while others who were born in the village grow up elsewhere; either way, Kathputli Nagar is considered home for them all.

4.5.7 LESSONS LEARNED

We found that although there is limited awareness about vaccinations in part because of public health campaigns, the overall understanding of the importance of immunization is limited. The residents did talk about the importance of the vaccination but also stated that it really is the job of the government to make care of it. They have bigger challenges like earning a livelihood and maintaining possession of the land, which they hope will get regularized one day. They agreed to the importance of immunization but could only name polio as one of the shots. This also reflected the lack of awareness and education of the masses from the government point of view.

There is a paucity of information on the need for timely immunizations. The task becomes more difficult when the earning member of the family is spending most of the day out in the streets and the people who stay at home do not have much authority or inclination to take any decisions especially that can anger the primary earner of the family. Being part of the census process unveiled some very interesting characteristics. The interest in the census process was lukewarm when the author first visited the community but the situation changed once a local doctor/community organizer was made part of the process. The presence of a community leader made a big difference to the kind of responses and information provided by community members. It can be attributed to the trust issues. The community understands the need to support each other for survival. Whenever they are faced with a situation where someone is asking them questions that they are unsure of, a safe route is adopted, which is to not divulge any information. During the first visit, when we visited without the community organizer, the most common response was "I do not know where he is and when he will come back, but he lives here." Once a trusted person is part of the process, the doubts can be alleviated and the residents felt a lot more comfortable engaging in a conversation. This became even better whenever the authors were able to hold a conversation with a group of individuals rather than one or two people. Being part of the census team also allowed us to experience the difficulty in identifying residents based on their officially recorded address. The residents knew who arrived into the community and who left for another city but to pinpoint each resident to a specific house was almost impossible. A main reason for this is the unorganized nature of the community. Every single interaction with government officials has to be considered for possible future implications like relocation to another locality, or making a case for electrical and water connections to the households. All of these factors contribute to the difficulty of applying regular information management systems, regardless of the purpose, to an unorganized community in an urban area. It also made us realize the need to rethink our approach and focus specifically on the children receiving vaccinations as the first and most important point of reference for creating a new kind of vaccination information management system.

The user profiles identified through ethnographic research enabled us to explore design opportunities for a vaccination information management system. Most of the critical immunization shots need to be administered to preschool age children, who spend the majority of their time around their mothers or grandparents in the vicinity of their homes. We saw the potential of creating reference points for the vaccination database that would be linked to the children needing the vaccination, their mothers or elders who are present in the neighborhood more than the father. There is a lack of accountability due to difficulties in tracking newborns as well as record keeping, making it hard to predict the requirements for vaccinations. This is more of an issue in unregulated urban residential communities (slums). A high percentage of the residents are in a constant state of migration, based on the availability of employment. They live where work is and the families move with them. Information management of vaccination or any other service for that matter relies on a point of reference, which in most cases is the physical address or some sort of identification (ID) card. The majority of the population that resides in the slums cannot provide that point of reference. The present system tries to plan for the demand based on information that is based on a flawed relationship, and this leads to very high levels of inefficiencies in the system. Any solution that aims at managing information about an aspect of such community life has to be based on the things the people do on a daily basis. At the same time the solution cannot introduce another step or a procedure in their daily routine. The community needs to understand, in their language, what are the benefits, both long term as well as short term. Last but not the least, any solution designed for such resourceful people has to focus on the positive aspects of their life and not point out what they lack financially.

4.6 OUR APPROACH—TRANSLATION OF LESSONS LEARNED INTO TECHNICAL OBJECT

It did not take us long to realize the futility of trying to adopt a standard address and ID card based system to work for this situation. We quickly realized the importance of creating a failsafe connection between the child and his/her vaccination records irrespective of their place of residence at a particular point in their life. We also realized the importance of creating a multi-layered system which would enable the community health care worker to find the child's vaccination records as well as update them in the most efficient manner possible. Furthermore, the challenge was not just limited to finding the right set of components that would complete the system but to also arrange it in a way that would inspire confidence in the members of the community as well as the community health care worker. Once we had some insights on the possible points of references inspired by the lifestyle in the community, the discussion quickly shifted to efficient modes of information management. Cell phones, tablets, cloud-based data management and stand-alone applications that would allow the community health care worker to create, retrieve, as well as update the vaccination information of a specific patient, irrespective of it being linked to a physical address or not.

Our solution was also driven by factors that are important within the context of India. India has the world's second largest telecommunication industry in terms of subscribers, and is the world's fastest growing in terms of new subscribers: urban communities experience wireless growth of about 7.37 million subscribers per month, while rural communities experience growth of about 4.04 million subscribers per month.[12] As of March 2012, the total number of subscribers in the country had reached over 951 million. Data plans in India are also some of the cheapest in the world, and mobile lines outnumber land lines at a 20:1 ratio. Mobile phone use as a tool for health care is not unprecedented; indeed, it is slowly becoming a popular choice for newly developing countries. mHealth,[13] the term coined as an umbrella description for mobile phone related healthcare applications, is popular in countries where the healthcare system is not as standardized and rigid as in the United States. ZMQ, a mobile application company, has created game applications that are non-platform specific for increasing AIDS and HIV awareness, as well as of other communicable diseases, in India. In addition to raising awareness, mobile applications have been developed to display results from CT and MRI scans to patients on products such as iPads or Android tablets. Encryption tools currently make it possible to store entire medical records in databases, though the practice has not been adopted well due to privacy concerns in many industrialized countries.

[12] Updated data available from the Telecom Regulatory Authority of India: `http://www.trai.gov.in/`; `http://articles.timesofindia.indiatimes.com/2012-08-02/telecom/32999709_1_subscriber-base-sectoral-regulator-trai-today-cent`

[13] `http://en.wikipedia.org/wiki/MHealth`

4.7 IMMUNE—A QR CODE BASED IMMUNIZATION APPLICATION

The approach that we are taking is two-part: a physical object that will store a Quick Response or QR code,[14] and a database application or app for a smart phone that is running an Android™ operating system.[15] Furthermore, QR codes are becoming increasingly common as a way to interact with static information. A QR code consists of black modules (square dots) arranged in a square pattern on a white background (see Figure 4.1) and the information encoded can be made up of four standardized kinds of data and this capability for data representation can be extended to virtually any data type through the use of supported extensions. The proposed solutions takes advantage of rapidly increasing cell phone penetration in Indian society by creating a smart phone-based interface that allows the community healthcare worker to create, access, and update a central immunization database. These tasks are accomplished by using a custom designed app which is highly customizable and scalable. The app allows for a visual data entry system and incorporates the picture of the child as part of the immunization record. The app creates a user-friendly and efficient way for the healthcare worker to do the job. This part of the solution, the IMMUNE app on a smart phone interacts with a QR code. This QR code is assigned to a record in the database at the time of creation of the record for the first time. It can be on a pendent worn by the child, can be part of a toy, it can be given to the father as part of an ID card or can be given to the mother and she keeps it in the family temple. The solution tries to create a non-prescriptive, flexible, and contextually relevant component that is the responsibility of the end user. They do not have to keep a health card that they cannot read; it is not required for them to make the information available to the health care worker in any specific format. To make the solution even better, if a QR code is lost, another one can be printed and given to the user.

4.8 TECHNICAL IMPLEMENTATION OF THE PROJECT

4.8.1 QR CODE HOLDER

In order to make the process as easy as possible for the healthcare workers, the QR code will be placed on an object that will be transported either to the doctor's office or kept at home for the healthcare worker's visits. The QR code will be housed inside a "unit" that can be applied to multiple objects. With the majority of vaccines being administered in the first 24 months, a unit that can safely be placed onto baby toys is ideal or the unit can just be carried with the parent on a keychain or bracelet. The unit should have magnetic ends so it can easily be removed from the bracelet and snapped onto other objects or hidden inside the objects or toys. The "unit" will ideally be made of a plastic or metal material. The bracelet will be very easy to "customize" and by adding materials and

[14] `http://en.wikipedia.org/wiki/QR_code`

[15] Although this idea was originally conceived by the authors, the development work was done in close coordination with student teams who often took the initiative at different stages of the project. The students were responsible for the technical development.

beads we believe the people will value their objects and QR code unit more and keep it safe. Also, beads can be added to the bracelet or unit to symbolize the vaccines the child has received.

Figure 4.4: Example of QR Code (`http://en.wikipedia.org/wiki/File:Wikipedia_mobile_en.svg`).

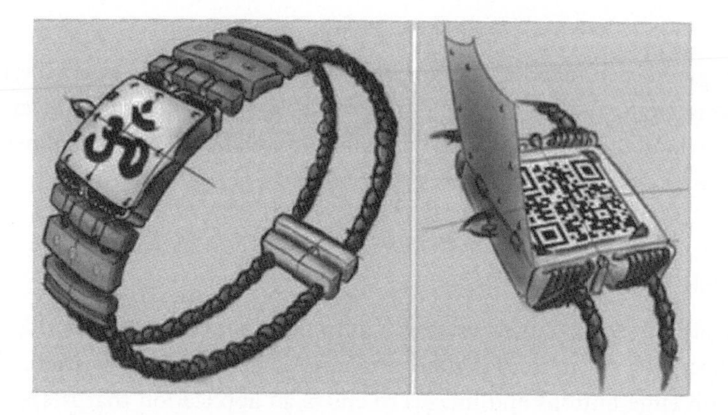

Figure 4.5: Early prototype of QR code based wrist band.

4.8.2 DATABASE

Presently, our storage system for holding child records contains the individual records that correspond to each individual who has been vaccinated. Ideally, a globally accessible database would be used by health care workers to find immunization records. The database stores a Child Record object,

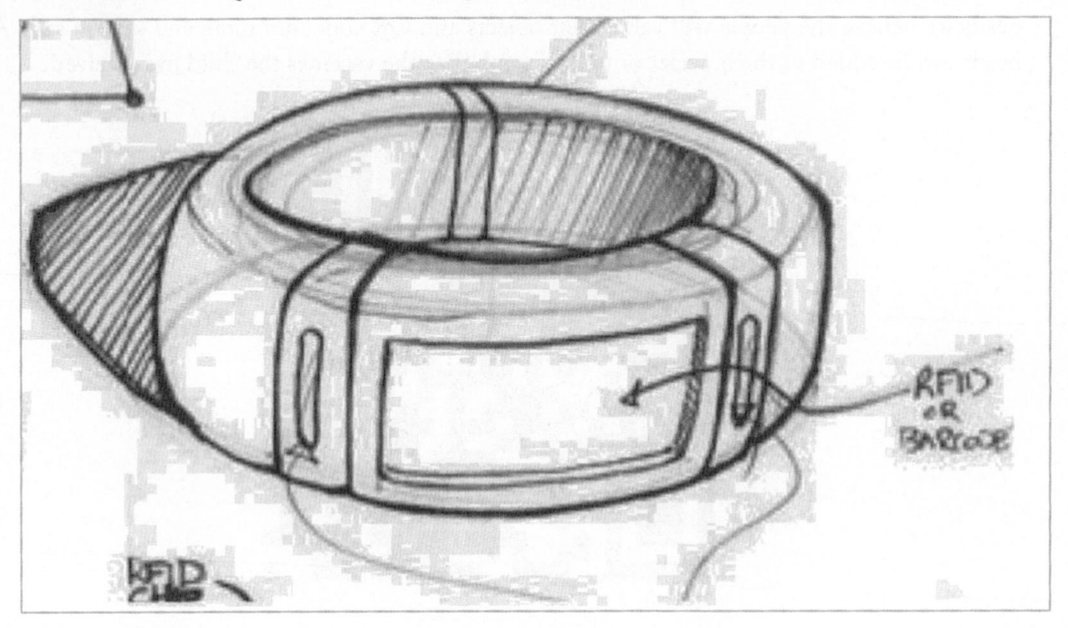

Figure 4.6: Initial idea using RFID or Barcode as an information storing mechanism.

which is unique to each individual. The Child Record stores a name, ID number that corresponds to an appropriate QR code, paternal and maternal names, date of birth, and an array of vaccines that have been previously given. The Android application will communicate directly with the database to allow access to any individual records as necessary.

4.8.3 USER INTERFACE

Since India has the second largest telecommunications network in the world, Android 2.3.3 will be used as the operating system running the immunization database application. The user interface will be simple with large buttons and images to create an application that is easy to use. The opening screen will be a set of four buttons, three of which are used for searching and the fourth for registering a new patient. One of the search functions will be to search via a QR code, which will open a QR scanner activity and read data from the scanner. All other search functions simply prompt the user to enter a name or an identification number to search the database with. The user interface must work closely with the database. The user interface will also be highly restrictive; that is, it will coax the user into entering a patient's data safely without affecting other patients' data.

4.8.4 DETAILS OF THE APPLICATION

Development of our application took part in two major phases—development of the database and development of the mobile phone application.

Development of the Database

The database that is used for our current prototype uses a text file to read and write from. The text file stores ChildRecord objects and is read when the class Database is instantiated. There are two classes that make up the foundation of the database: the ChildRecord class and the Database class.

ChildRecord Class Each ChildRecord object contains strings for a child's name and parents, integers for their birth date, and a Boolean array that stores vaccination data. Each of these objects has a corresponding getter method, so that they can be accessed easily by external classes and methods. The constructor for a ChildRecord object takes seven parameters, as shown. In addition to getters and setters, the ChildRecord class also has a `print()` method and a `generateID()` function. These methods print out the record to standard output and generate a unique identification number for a child, respectively.

Database Class The Database class consists of an internal array of ChildRecords, which is built from a text file. When the database is modified, it will write to the text file using the function `addRecordToDataBase(ChildRecord c)`. There are also methods to read in the stored text values and convert them to a Boolean array.

```
{
    name = s;
    maternalName = m;
    paternalName = d;
    birthYear = y;
    birthMonth = mo;
    birthDay = da;
    ID = generateID(mo,da);
    vaccines = a;
}
```

Figure 4.7: The internal function of the constructor for a ChildRecord.

Development of the Application

The second phase of the project consisted of the development of the user interface and the constant communication between the model and the view. The application includes an icon to display on

```
Intent thisIntent = new Intent(SearchByNameActivity.this,
    PatientDataActivity.class);
thisIntent.putExtra("personName", person);

startActivityForResult(thisIntent, 1);
```

Figure 4.8: Sample code for how to start a new activity and wait for result in Android.

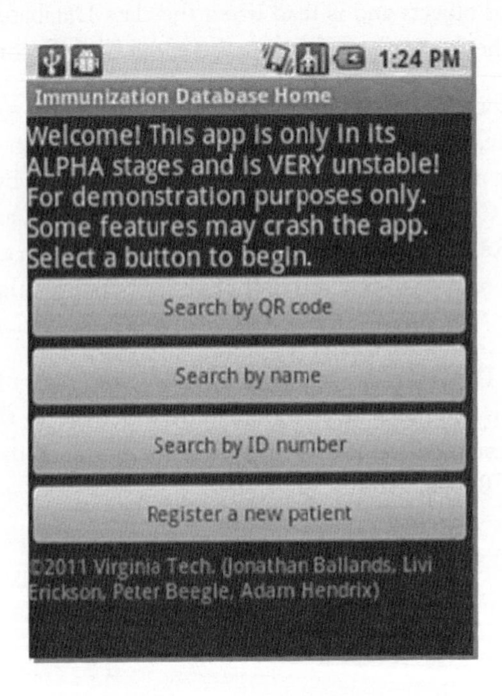

Figure 4.9: Home screen of the application.

the home screen once installed. Once the application has started, the user is welcomed to a screen consisting of four buttons. This screen is formatted in the main.xml of the application under its Resources folder. The user can select any of the buttons. Each button will start a new activity in which the home screen will wait for a result.

In this code, a new "intent" is created, which includes the class the intent is coming from, and the class the intent should access. This particular intent is created in the SearchByNameActivity class, so an "extra" is put into the intent that contains the name of the person-of-interest. This allows the PatientDataActivity access to the name. Finally, the method `startActivityForResult()` is called. This method sends the intent to the next activity with a request code. This request code is important, since when the intent comes back, Android has to know where the intent came from. The integer 1 lets the class know that the intent came from the PatientDataActivity and not elsewhere.

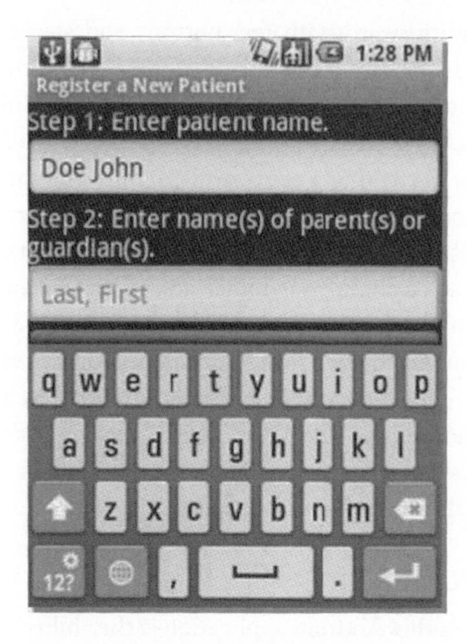

Figure 4.10: Screen for registering a new patient.

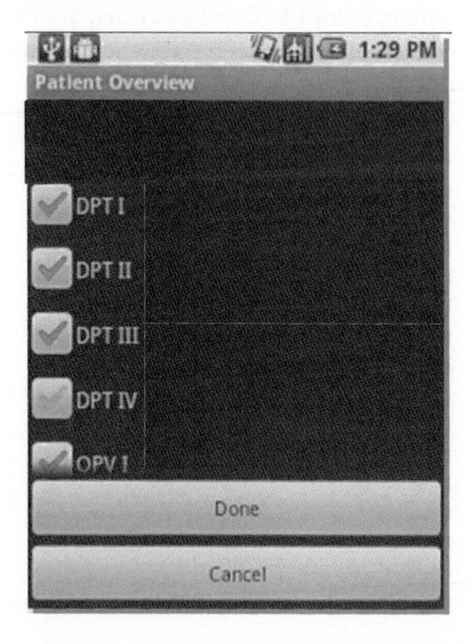

Figure 4.11: A healthcare worker can easily mark or see administered vaccines on this screen.

The entire GUI is straightforward with few buttons and step-by-step instructions, making the application easier to use. If the user wants to add a new patient, this screen will appear. Notice how each text entry box is accompanied by an instruction. There are only three steps to creating a new patient; everything else is done internally (see *Development of the Database*). A pre-existing patient is just as easy to search for as creating a new one, with only one step: typing the name of the patient and pressing "Search." Once the user has found the patient they are interested in, they will be shown the following screen. This is where the user can modify a patient's data and save any changes. The entire page is scrollable, so stroking your finger across the screen will cause the data to move up and down. In this particular screen, a dummy patient "Jane Doe" is active with her date of birth is displayed directly below her name. Notice how some of the vaccinations already have a green check mark beside them. This indicates that the vaccination has already been administered. If the healthcare work were to administer "DPT IV," they would simply tap the check mark beside "DPT IV" and press "Done" to save their changes. If unwanted changes are made, simply tap "Cancel" to discard all of the previously made changes and restart.

4.8.5 USING QR-CODES

One of the main features of this Android application is the ability to use QR-codes to find a patient. If the user chooses to search by the patient's QR-code, the following screen will appear.

The user must simply position the phone to scan the QR-code. This is accomplished through the user of open-source software known as Zebra Crossing, or ZXing. Once the QR-code has been scanned, the user will be taken to the Patient Overview screen if patient was found in the database.

4.9 SECOND GENERATION PROTOTYPE OF THE APPLICATION

Figure 4.12: Second generation prototype for the IMMUNE application.

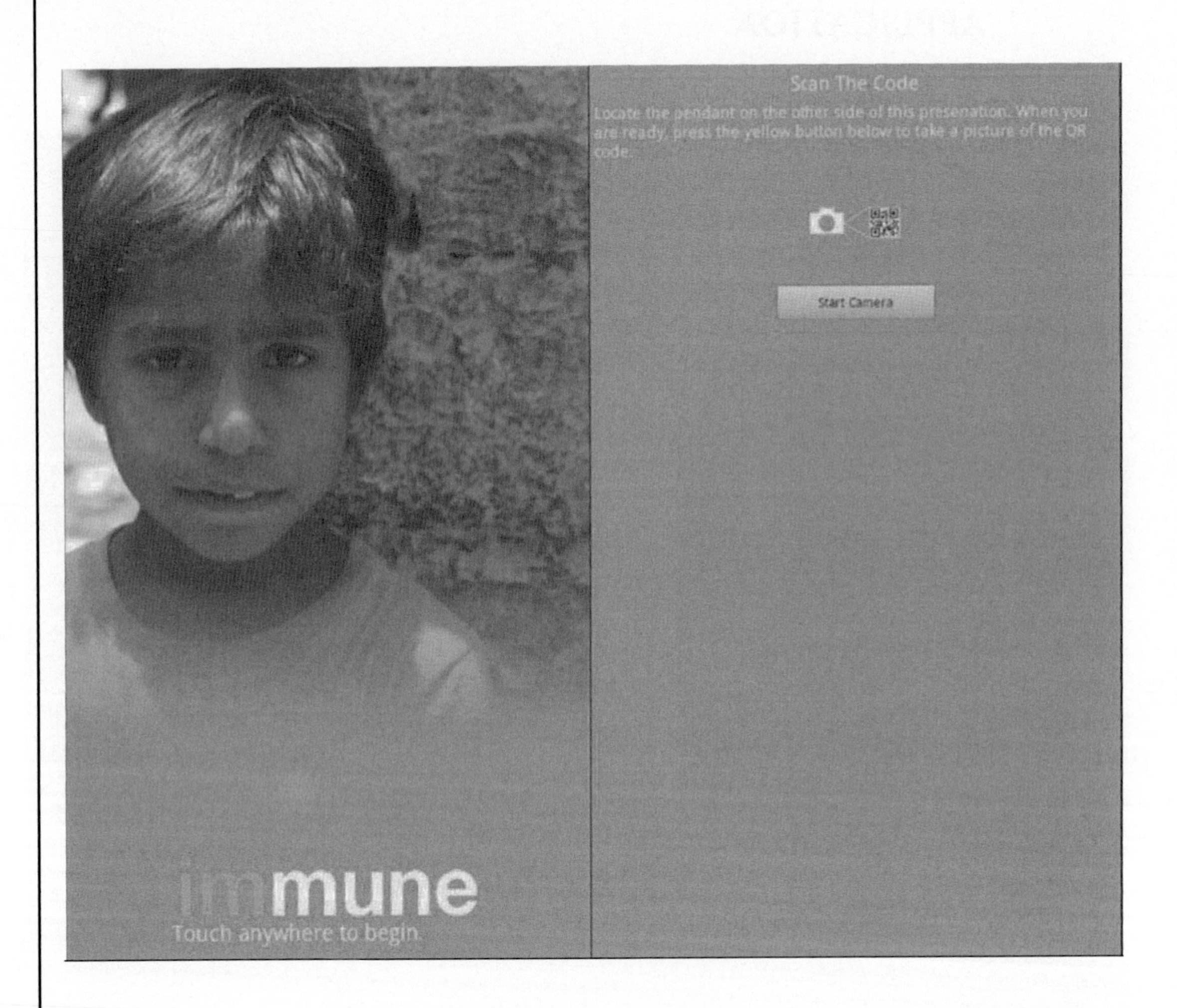

Figure 4.13: A series of four screenshots from the second generation IMMUNE prototype with similar functionality as before but more refined user interface. *Continues.*

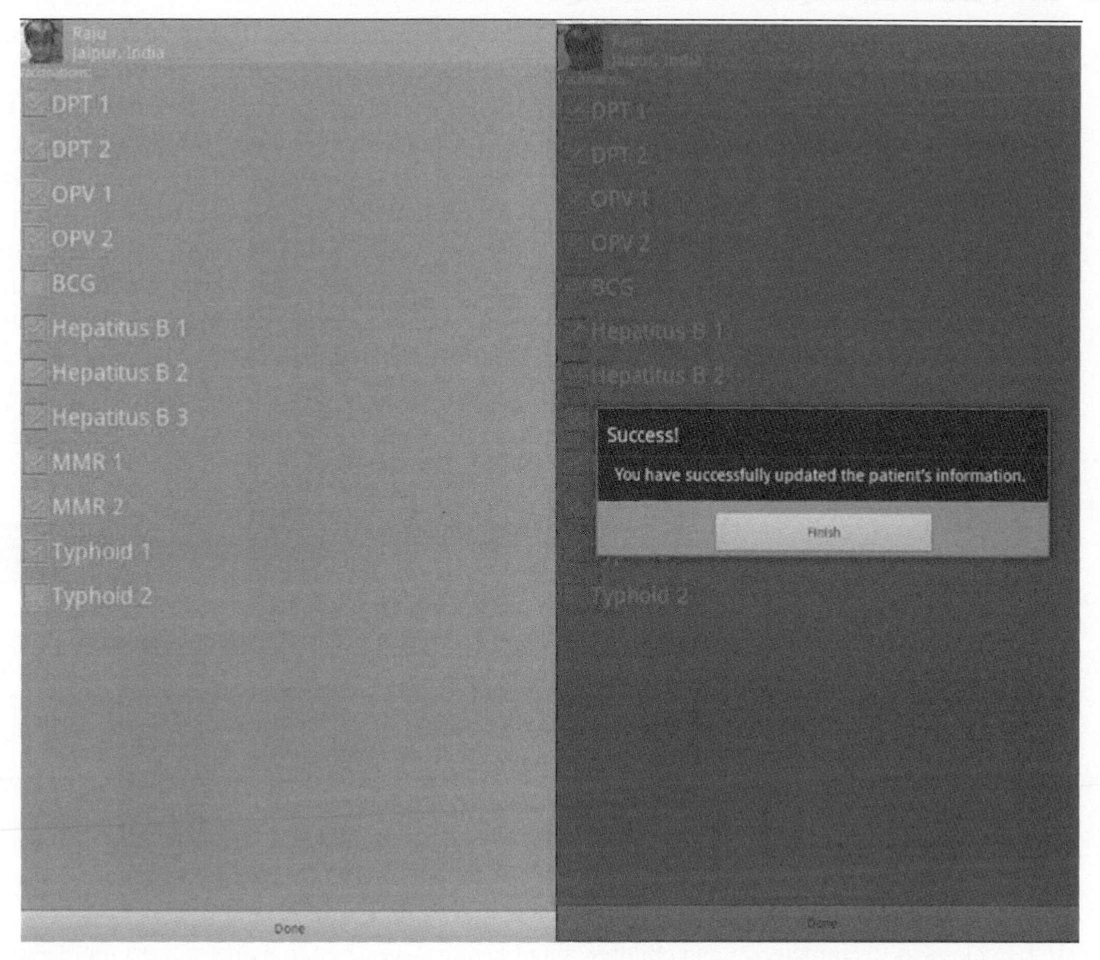

Figure 4.13: *Continued.* A series of four screenshots from the second generation IMMUNE prototype with similar functionality as before but more refined user interface.

4.10 REFLECTING ON SOCIOMATERIAL INFRASTRUCTURE

We realize that the infrastructure we have built is essential to any design effort and especially to an effort that involves teaching and training of students. Being cognizant of the needs of such an effort is important and so is the effort to put all the pieces in place but be open to change. Within this theoretical framework that guided our practices, we took many other steps as we discussed earlier—we built partnerships (across engineering and industrial design faculty and with NGOs in the field), and undertook a series of activities, including: independent study, incorporation of design projects in a large freshmen course, summer REU (Research Experiences for Undergraduates) program, and a full-fledged class offering. We identified the importance of picking a few key projects and clients and working on those projects and with those client partners for a longer time period. This infrastructure building is essential and allows for incorporation of mentors, more buy-in from the clients, improvement of the design and product over time, while providing enough variations to keep the students interested. Therefore, within our work, sociomateriality plays a critical role by helping us overcome an inherent dualism present in learning and design practices. Overall, the immunization application we designed—IMMUNE—addresses some of the most labor intensive and culturally insensitive aspects of the current vaccination record keeping system and provides an alternative without creating a new approach from the ground up. It has resulted in a short learning curve and high acceptance of the idea.

Table 4.6: Capable and Convivial Design (CCD) Framework Applied to the Immunization Design Case Study

CCD Principles	QR Code Immunization Design
Ease of Accessibility	Commonly used platform as most phones in India have a camera; QR codes are cheap to produce and replace; allows for sharing of information across users
Expressive Creativity	The use of different kinds of artifacts on which QR codes can be placed allows personalization of information
Relational Interactivity	Involvement of children, parents into the health care process; formation of relationship among health workers and the beneficiaries
Ecological Reciprocity	Creation of longer term health practices; education of parents and children about health issues; health workers learn more and understand the context better through close involvement

4.11 SPINOFF PROJECT

The IMMUNE application has already been appropriated for other projects such as an application for helping an organization that builds and supplies artificial limbs to handicapped patients.

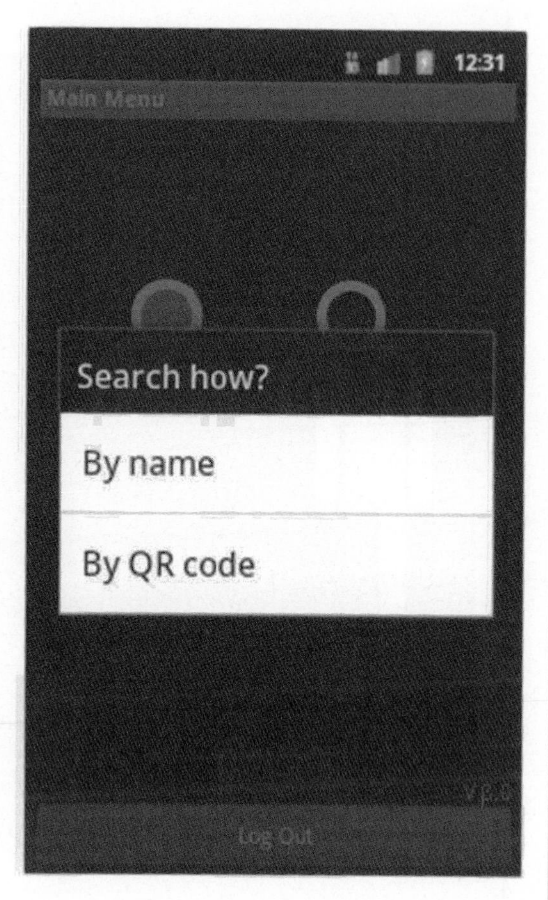

Figure 4.14: A series of four screenshots depicting an Android application for registering new handicapped patients. *Continues.*

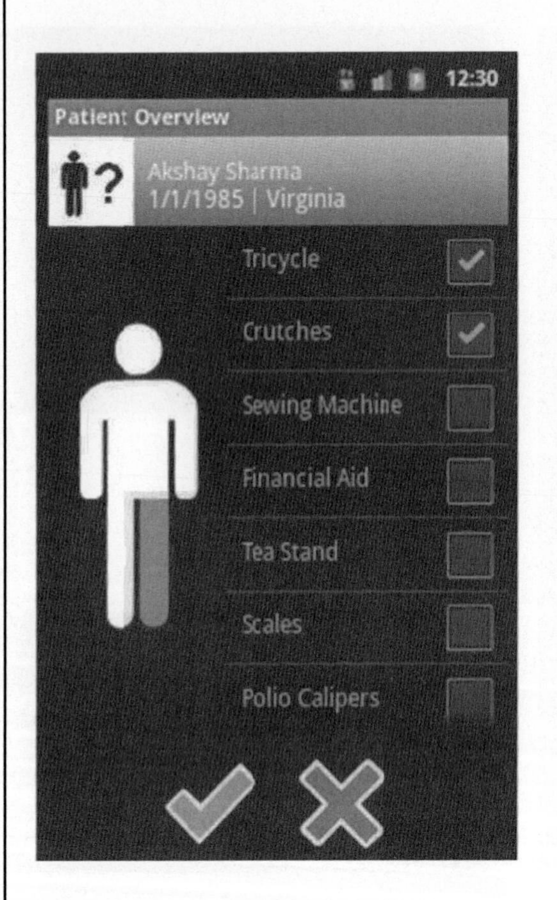

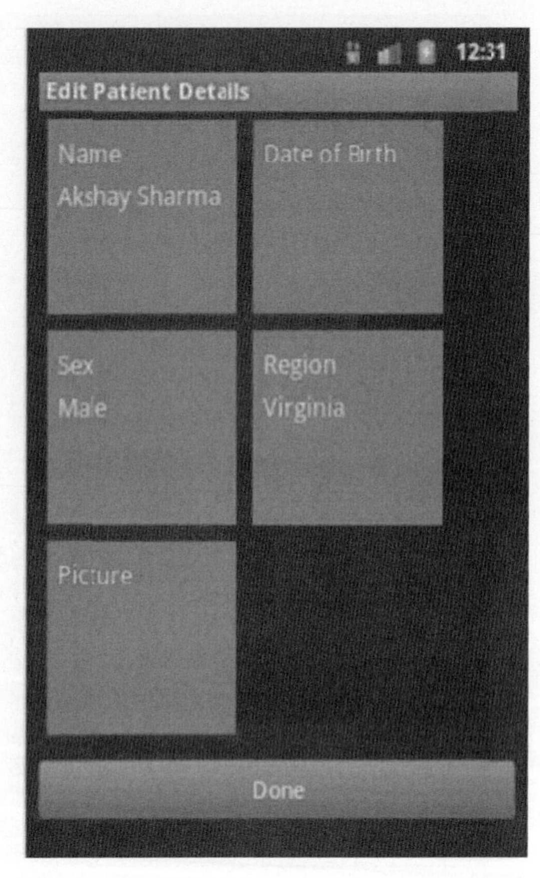

Figure 4.14: *Continued.* A series of four screenshots depicting an Android application for registering new handicapped patients.

4.12 FUTURE WORK

In India, because a large portion of the population is not aware of the importance of immunization, we hope we can keep developing our design to optimize the amount of users. This could be through making our application available to other phones that are currently not compatible with our application. With feedback from various users, health care workers, and users as a whole, the application will be tweaked and updated to become more efficient, able to store a large amount of information, and eventually be available in different languages. The design of the toy holding the QR code may or may not be altering depending on how the children keep a hold of the item. If this idea does not work, we will work on new designs and their implementation.

CHAPTER 5

Design for Development Course and Outreach Initiative

In this chapter we describe in detail a course that we have developed as part of our overall program. This course brings together different elements of our education and outreach program and showcases one way in which a program such as this can be instantiated within an educational institution. The objective behind using a design-based course was to engage students with global issues in a substantial way so that they could work on projects that made a difference. The ability to work in a global world has emerged as the foremost skill that needs to be developed among engineers. Policy makers, academics, and practitioners all acknowledge that global collaboration is essential for sustained economic progress and for solving critical social and environmental problems. Over the past decade NSF, academia, and industry have invested significant resources to identify critical skills required by engineering students and practitioners to succeed in the 21st century. Informed by these findings an increasing number of research programs and institutions are devoting significant resources to prepare engineers for a global world. Although institutions continue to push this agenda forward, they are constrained in their efforts by the absence of comprehensive knowledge on global engineering work that can inform their pedagogical efforts. Furthermore, efforts such as study abroad are common but not highly subscribed. A very small percentage of the student population participates in these experiences and universities are faced with coming up with innovative ways to teach students global skills.

5.1 DESCRIPTION OF THE COURSE

We offered an interdisciplinary course open to industrial design, architecture, and engineering students in fall 2011.[1] The course was titled "Design for Bottom of Pyramid" in the industrial design and architecture department and "Engineering Design for Social Development" in engineering. The class duration was ten weeks with two meetings each week. It was advised by three faculty members representing industrial design, engineering, and human-computer interaction. The course attracted

[1]Johri had earlier offered a course on global engineering work to engineering students and realized that to really engage his undergraduate engineering students it was essential to provide them design experiences as most students found reading and case studies difficult to grasp given their learning habits of solving close-ended problems. Students were not used to ill-structured and open-ended problems, a staple of the real world, and therefore this project was devised to present them authentic design problems. Furthermore, given the interdisciplinary nature of work in the real world, a concrete decision was made to make the course and the projects interdisciplinary that required genuine interdependence among students from different majors.

an even mix of students from engineering and design, and the class was divided into five teams of four students each.

Figure 5.1: The authors (Sharma on the front left and Johri in the front center) teaching the interdisciplinary course.

5.1.1 COURSE OBJECTIVES

Technologies such as the Internet and mobile phones are already playing a major role in economic development, learning, and social empowerment in developing countries. What might be the impact if that technology was specifically designed for users in these countries? The primary objective of the course was to address this question by developing and implementing a studio style, project-based course. The course will bring together students from various disciplines who are interested in developing computational devices and applications for people in sub-Saharan Africa and India. Collaborating with area NGOs and immigrant communities, students will conduct interviews to identify needs, generate concepts, create prototypes, and test their use. The emphasis of the course will be on creating new designs through understanding and empathy with the people and settings for which the innovations are intended. The semester-long class will result in a collection of working technology prototypes that we will share with the Virginia Tech community during an end-of-semester "critique." A sample of projects is listed in Table 5.1. We also invited guests who are leaders in an emerging interdisciplinary research area called Human-Computer Interaction for Development (HCI4D) to evaluate the students' projects and to help us determine which ones merit further development.

5.1.2 PARTICIPATING FACULTY AND DEPARTMENTS

The course was a collaborative effort between Human-Computer Interaction (HCI) faculty in Virginia Tech's Computer Science Department, faculty in the School of Architecture and Design's Industrial Design program, and faculty in the department of Engineering Education.

5.1.3 TIMELINE

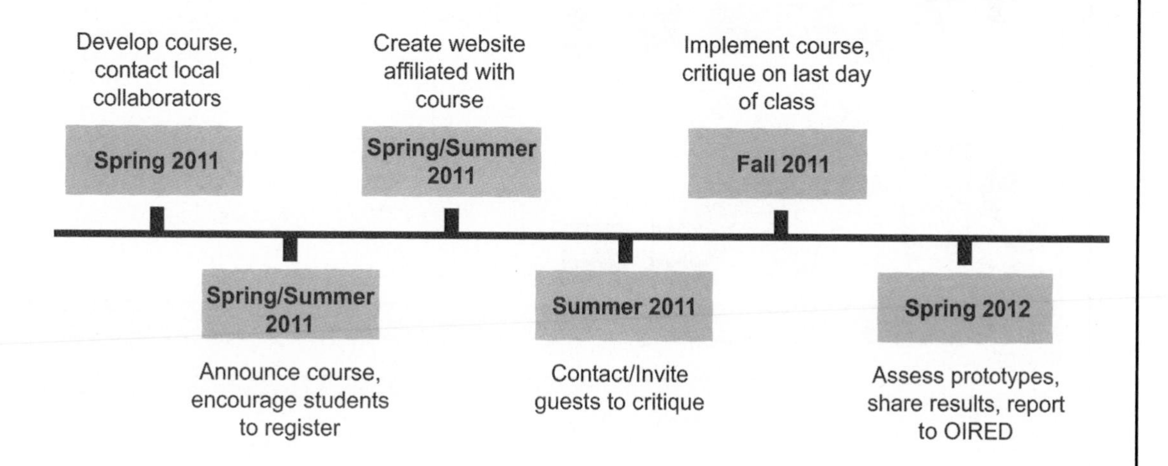

Figure 5.2: Timeline.

5.1.4 EVALUATION OF OUTCOMES

There were two phases of evaluation for our projects that were done in the studio. First, students filled out a course evaluation survey at the end. Second, during the final critique, students' prototypes were assessed, and demonstration videos were created to be uploaded on the website. The proposals created the next step in the process so that students enrolled in the course in subsequent semesters could develop the ideas further.

Table 5.1: Sample Projects from the Course *Continues.*

Sample Projects from the Course
Project: Jaipur Foot
Purpose: To create a system that can reliably measure foot sizes for handicapped patients and create a database on sizes that can be used to create artificial limbs with more precision. In addition, create an electronic database system that can be easily updated.
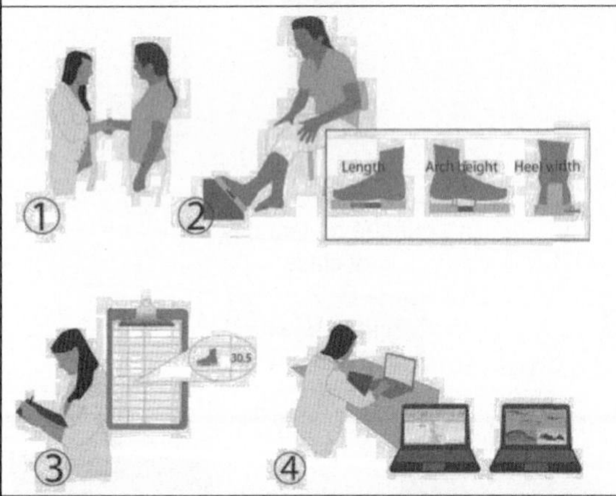
Project: Push Cart
Purpose: Design a more easy to use push cart for vegetable and food vendors. The cart should be able to keep the produce fresh.

Table 5.1: *Continued.* Sample Projects from the Course *Continues.*

Sample Projects from the Course
Project: Mobile Learning Device
Purpose: Design a device that has a computing and projecting capability and can be easily carried for teaching in rural areas. The device should be rugged and battery powered without requiring electrical power sources at the point of use.
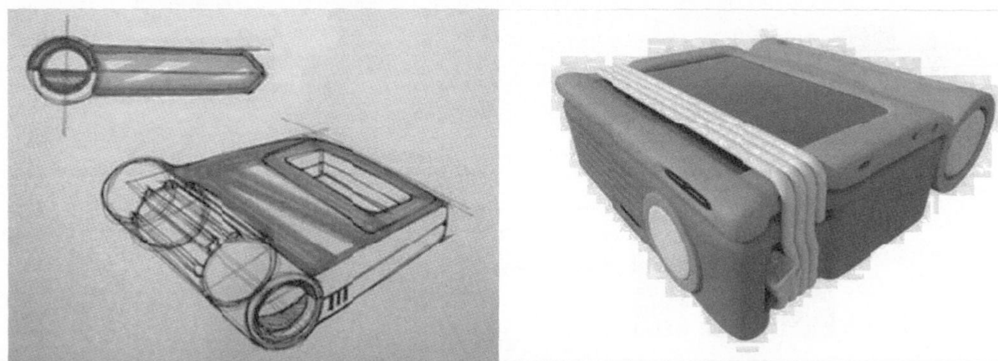
Project: Cell phone charger
Purpose: Design a non-electric charger for mobile phones for use in developing countries.

Table 5.1: *Continued.* Sample Projects from the Course

Sample Projects from the Course
Project: Financial Literacy
Purpose: To design a financial literacy system to teach compound interest to semi/illiterate women in rural India engaged in micro-financing.
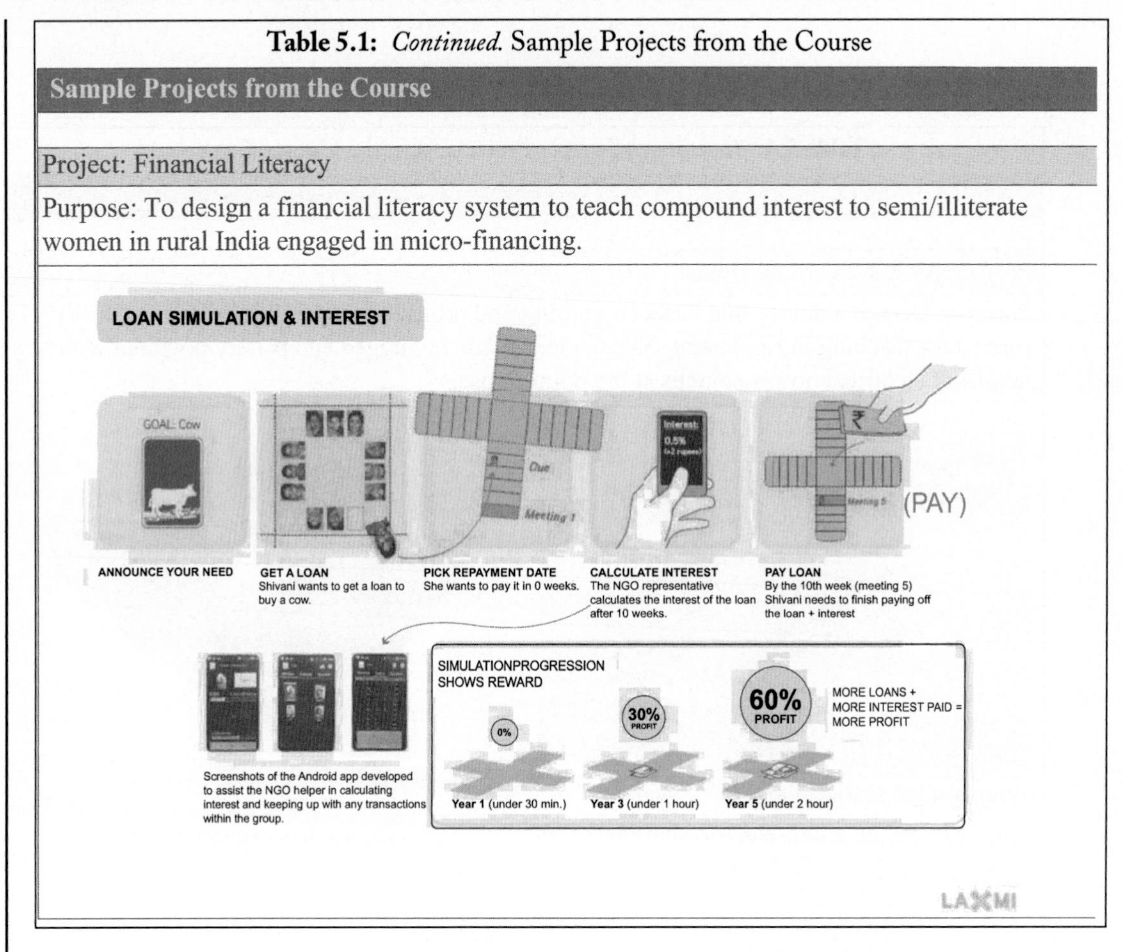

5.2 COURSE ASSESSMENT

Toward the end of the semester quantitative and qualitative data were collected from students using focus groups and surveys to assess the course offering. The research protocol was reviewed through the institutional review board, and participants consented to their participation.

5.2.1 FOCUS GROUPS

Two groups of six students each participated in the focus group near the end of the semester, after they had presented their final projects but not submitted their final project report. Each focus group lasted between 30-45 minutes. The focus group protocol was open-ended and started by asking all participants to write down whatever came to their mind when they thought about the course [Morgan, 1997]. The objective of this exercise was to prevent groupthink and ensure that each student made time to write down their thoughts. Students were then asked to read aloud a few of the issues they wrote down, which was used as a starting point for the focus group; subsequently, the written notes were collected at the end of the focus group for further analysis. During the focus group each participant was given enough time to express their thoughts, and participants were called on to ensure equality of participation. Overall, the participants expressed their thoughts openly and were not hesitant to critique the class.

The findings from the focus groups highlighted many interesting issues. From the outset, students expressed positive feelings toward many aspect of the class such as its interdisciplinary nature, open-ended but real-world projects, the ability to spend time on the project and do in-depth work, and the ability to positively impact someone's life in a developing country (in this course, India and Kenya were the target countries). All students agreed that they were really motivated by the opportunity to work on projects that would be field tested and that were identified as real needs of users as one participant commented, "We all have some interest in social development. We have the blessed opportunity to go to a school like this but it is important to realize that something as small as charging a mobile phone can change somebody's complete day." They also suggested that since all of them had self-selected to take this course, they were all equally motivated (except a couple of students). They further expressed their opinion that the class would not have been as effective if it was a required course, as real interest and passion were needed to labor through many difficult issues that emerged as they worked on their projects.

In addition to a prosocial motivation, students also expressed other motivations for taking the course. First and foremost was the ability to work with students from other disciplines. The course was open to students in the industrial design and architecture programs as well as all engineering disciplines, including computer science. It attracted students from six different disciplines. Each team had at least one student from design/architecture and one student from an engineering discipline. Students reported that they learned that everyone thinks differently and the engineering students reported that they realized that not all that was technically feasible could be designed into a usable product and the design students reported that they realized that just the design of form does not mean technical feasibility. Interestingly, computer science students reported that they learned a lot even from working with other students in their major as they rarely got a chance to collaborate this closely on a project as most of their class assignments were more competitive in nature (each student had to solve the same problem). The students also commented on the nature of problems as a motivator beyond their content per se. As one student remarked, "Given the context, the level of accessibility is easy since the projects haven't reached a level of complexity where you cannot really contribute, here you can actually contribute." Another student added that she thought that what made the class unique was the fact that a connection with a field site was already made for the students.

When asked specifically about what they learned, in addition to social development issues, students commented on many other aspects of the course showing that the learning extends beyond a specific issue and can be transferable. One student remarked, "I learned the importance of having people from different majors to do design. All I usually do in my mechanical engineering classes is work with other mechanical engineers and we all think similarly." Another student from industrial design expressed her learning succinctly by stating that she learned to "try to think of a problem from someone else's perspective." She further stated that she formed an understanding to keep in mind to be very aware of how different people you are designing for are and check everything before moving forward as it might be completely wrong. Another student gave an example of how they were trying to use green and red buttons to designate going forward or stopping in an interface they had designed before she realized that when she had visited the site in India she had not seen a single traffic light. A similar example was given by a student from a team that designed a bicycle-powered phone charger. He said that they realized that the charger will have to engage with the bike against the chain as opposed to the tires as the tires got damaged quickly and the riders could not afford to change tires frequently. Therefore, students learned about taking context into the equation and also thinking about different business models or financially viable solutions. The solar charger was designed as a "station" for charging multiple phones so that it could be a profitable venture for a small shop owner.

Students gave specific examples of how the project in this class overlapped with other courses and helped them improve their outcomes in other classes. For instance, once student who was working on the design of mobile phone software talked about his experience working on the Android platform. He said that he leveraged that for other classes and by using the same platform across assignments he was able to learn things a lot quicker. He said that he realized that the abstractness of computer science, particularly of software design, makes it harder to share and show what one has done and therefore he realized that in addition to documentation one needs to trust the other person and their skills. Students also noted that they were faced with many choices and options, even when working on a limited design problem, and they learned to deal with ambiguity and design for appropriation of the design by the users. For instance, a design student working on a QR code artifact interface said that after experimenting with many options he realized that the best option is to design it so that the users can put the QR code on any artifact they prefer.

Finally, students commented on their positive team work experience. Barring a couple of students, all students engaged highly with their team members. There was good communication between the groups as well as with mentors at the field sites in India and Kenya. There were some issues with communicating with field mentors, but they were resolved by engaging through different communication media. Students mentioned that interdisciplinary team work enforced creativity as brainstorming among students was common. Bigger and better ideas came from their group work and slowly they learned about the expertise of each member which helped them maximize their productivity. Overall, similar responses were also reflected in a survey administered subsequent to the focus group interviews.

5.2.2 SURVEY RESULTS

The students were asked to take a survey about their course experiences subsequent to the submission of their final project report. The survey consisted of 10 items that assessed various aspects of the students' experiences. The primary purpose of the survey was to triangulate the data acquired through the focus group interviews and to give students another avenue to talk about the shortcomings of the course. Results are reported in Tables 5.2 and 5.3.

The survey responses shed more light on student experiences with the course and also provided some quantitative data on student learning. We plan to do a follow-up of student assessment next semester to assess students' transfer of learning across courses. We also plan to analyze student design solutions from early in the semester to later designs to see the development of ideas over time.

5.3 OUTREACH COMPONENT

Our initial ethnographic work provided valuable information but also helped us in forming partnerships on the ground. We realized early on that it is one thing to have passion for "doing good" but a whole other story to actually take it to fruition. For that, you need to know the right people—those who can help at important junctions in your journey. For us this has meant meeting with people who work at the community level with the end users as well as experts in the policymaking areas of social work. We made connections with NGOs, private firms, governmental agencies, and academic institutions in India.

It is very easy to do all the hard work and collaborate with someone with a different agenda and fail in the implementation stage. Over the past three years we have traveled to India consistently in order to form new partnerships but also to sustain existing partnerships. These are helpful to use in many ways. For instance, while students in the U.S. were working on class projects we were able to introduce them to experts in India and they could contact these experts as they worked on their design projects. They were able to have Skype calls or phone calls and get quick feedback on their projects. In turn, this interaction assured that our goal of producing something useful for our outreach partners was sustained. Our work and the needs of our partners were always aligned through interactions like these.

In addition, during out trips to India both the faculty and the students are able to work more directly with our partners. We are able to act as intermediaries between the end users and members of the NGOs. The students are able to work together with children on small learning tasks—such as drawing and coloring a world map—and thereby both experience interaction with someone from another context and in turn provide the same experience to others. These kinds of small experiences are invaluable when it comes to learning more about others.

On certain occasions we have taken a more direct approach of helping our partners such as the design of a projector-based table device as an alternative to a heavily constructed computer and project system—a project we undertook with one of our partners. This project helped us understand the difficulty in diffusing an innovation and the path dependency that exists in the way government services are provided. Even though the product we designed was a slimmer and faster version the heavier version continued to be implemented as public schools had signed up to receive that device through a tender process and to change that was almost impossible.

Table 5.2: Student Response to Open-ended Survey Questions *Continues.*

Student response to open-ended survey questions	
Open-ended Survey Question	Student Responses
What are some things you learned through your participation in this class?	I learned the importance of testing out what is already on the market because in doing so you can learn a lot about the current issues in the design. I also learned not to try and think of the best solution to a problem immediately, but instead to do background research and think about any and all possible options. After coming up with a number of options, I then learned the importance of moving on to the next part of the project and not getting caught up on the selection process or any other part.
	I learned how important it is to understand the many different view points on a particular problem. Working with people from different disciplines not related to engineering also helped our design work much better than if we did not. Information is hard to find on some topics, but most everything has something online. I also learned the basic elements of the design process.
	I developed technical skills related to the Android platform (Android Fragments, ActionBar, ViewPagers, Photoshop Actions). I also learned how industrial design students generally go about designing. I learned about microfinancing and about requirements analysis.
	I learned about difference in problems across areas, such as in America how different Immunization is from India. As well, how different majors can join in to design something truly of purpose.
	I learned a lot about teamwork in programming, which was something I had been unfamiliar with. I also learned a lot about the medical system in India and I thought it was interesting that we had the ability to work on a real life solution for it.
	Though having taken this class before as well, it was interesting to work with a computer science student one and one. There was a lot of back and forth dialogue to get to what was implementable in the given amount of time. I also learned of problems in other parts of the world- mostly the Kenya project- and how that is something that I never would have even thought of or imagined being a problem (lack of power/electricity shortage).
	I learned about how challenging it is to design for social development but also how helpful it really could be in developing countries. I also saw the importance of working in teams, especially with people from different backgrounds. I now understand the importance of making several different models/prototypes to get a better idea of what a design is and whether or not it will work.
	Working in interdisciplinary teams helps balance creativity and practicality. Project designs perform their best when they meet the maximum number of constrains, like financial and social constraints for example.
	Always allot double the amount of time you originally estimate the completion of something. It is far more difficult to design for a different culture than I had thought.

Table 5.2: *Continued.* Student Response to Open-ended Survey Questions

The most FRUSTRATING aspect of this course was:	I didn't really think there was anything that was particularly frustrating, but if I had to choose one thing it would be trying to think of a solution that would best benefit the Kenyans because we aren't there observing their day-to-day lives so we just had to create something based off of what we knew from our research and interview (with mentor in Kenya).
	It was frustrating when some of our group members didn't show up on time or even at all, but that's not really the course's fault.
	Having to work with people who just don't care
	Not enough time to work on projects.
	Trying to understand an NGO member's accent while in a Skype call with her. Email would have been better.
	Being a mechanical engineer major, my topic was mainly focused with programming, which was hard to join in on.
	I think that the nature of our project made it difficult for us to always be working evenly. For example, the research phase was well split up, but when it came to our actual project development it was a little unbalanced because there were some days where I would be coding non-stop and other days when I didn't have much to do at all.
	Coming up with a mutual idea we could all agree on to get the project rolling took waaay too long to describe SHG and get a good firm understanding on it in order to design something for it. Did not help when some members would read the material online and others would not. Therefore when in class the next day have to back track to get the other member caught up to the rest of the group just to begin throwing out ideas.
	The inability to test our final design
	It was tough learning to work with a computer program who thinks so differently than me.
	I really didn't have too many complaints for this class besides my own inability to fully focus for 2 hours in the morning.
	Low interaction with the rest of the groups. More information/teaching about how to design for developing countries and important considerations. Direction for further reading on the subject.
	Getting stuck, sometimes, at one point for a while. Yet, having the professors pushing us to the next step helped our team to move on.
	Confusion. It was not until over halfway through the course that we finally had a concrete direction. Too many weeks were spent trying to decipher the real intent of the project.
	Being in a group with a project dedicated to android programming with only one computer science student. We were limited in our time capabilities for what could be done in android. This left three ID students to design basic user interfaces, as well as plenty of research. Sometimes it felt like there was not enough work for the ID students.

Table 5.2: *Continued.* Student Response to Open-ended Survey Questions

The most INTERESTING aspect of this course was:	Being able to go through all of the different steps in the design process all the way up to developing a working prototype. Also, the amount of freedom we had in doing so and being able to work on a "real-world" issue that would actually benefit someone somewhere.
	The fact that we got to work with other people from different disciplines and work on real world problems.
	Being able to think for yourself
	Fitting in with a pre-existing design. Practicing everything. Actually helping people.
	Seeing how different groups interacted, and the final design of each project.
	Being able to work on a project that was not severely restricted in terms of deliverables and grade scale.
	Seeing how all the projects grew and where they ended up. Hope to get feedback on the projects and see where they go.
	Being able to work in teams of different disciplines
	Creating a working product in the end.
	I like the concept of tackling real world problems for countries of lesser development, because the problems weren't too complex to find a solution. Also, the fact that our solution could be implemented and make a difference in an entire community is very satisfying.
	Working in interdisciplinary teams. It was also good to have predetermined projects with goals. Having three professors that are passionate about the course.
	Seeing how real life problems could be solved and how people from different fields could work together to solve them.
	Actually working on something that has the opportunity to make a difference, or at the very least elicit further discussion into matters that are often overlooked.

Table 5.3: Student Response to Likert-scale Survey Questions about the Course

Student response to Likert-scale survey questions about the course		
This course ______ my interested in social issues in developing countries:	My understanding of design increased as a result of the hands-on projects:	I found it useful to interact with someone in India and/or Kenya:
Greatly increased 10 (63%)	Strongly agree 10 (63%)	Strongly agree 11 (69%)
Partially increased 6 (38%)	Agree 5 (31%) Neutral 1 (6%)	Agree 3 (19%) Neutral 2 (13%)
Neither increased nor decreased 0 (0%)	Disagree 0 (0%) Strongly disagree 0 (0%)	Disagree 0 (0%) Strongly disagree 0 (0%)
Partially decreased 0 (0%) Greatly decreased 0 (0%)		

Figure 5.3: Prof. Sharma at a SHG meeting.

Figure 5.4: A Virginia Tech student working with students on drawing and painting a world map as part of the outreach initiative in Rajasthan, India.

Figure 5.5: The authors at Barefoot College.

CHAPTER 6

Conclusion–Lessons Learned

A program of this nature requires significant resources. It also takes time to put into place the infrastructure required to engage with such projects. Therefore, one of the first caveats we wish to bring to the attention of those considering such courses is to be certain that they are willing to commit to it seriously. They should also plan for the long term, as short term efforts are not likely to lead to success. The downsides of a failed effort are many, including no learning experience for the student and actually harming the partners on the ground. But once the motivation is there, here are some specific issues we have synthesized from our experiences that can help improve the learning outcomes:

1. **Form long term partnerships on the ground**

   It is essential to build long term partnerships on the ground and initiate engagement with stakeholders with the intent of building a long term partnership. Quite often socially leaning projects are undertaken impulsively because the instructor or students are excited about an issue or problem. Although prosocial motivation is a critical factor, the hurdles and barriers faced in actual implementation often lead to swift demotivation. As we discussed, we had to overcome immense challenges and found that our long term vision sustained us through the lean periods. A long term partnership also ensures that projects are tested in the field and there is a feedback loop between the work done and its impact. This, we have found, serves as additional motivation for students. Since students engage in numerous design projects throughout their curriculum, most of which are standalone problems with little connection to actual practices, the opportunity to work on implementable projects is highly desired by students.

2. **Conduct on-the-ground needs assessment**

   A second issue, related to the first, is to ensure that the needs assessment leverages the context of the problem. Often the tendency is to define the problem first and then look for appropriate settings or to define but not refine the problem once a setting is selected. More often than not this approach results in interesting designs with little feasibility or future. As we discussed in the case study, we carry out an extensive needs assessment. Given our partnerships on the ground and familiarity with the context, most of this stage consists of primary research. This need not be the only source though. There are numerous secondary sources available and their numbers are increasing each day. For instance, we have found extraordinary resources through the United Nations and World Bank website but also on commercial sites such as YouTube

™. Citizens post videos regularly that can form an important resource to understand what the targeted population does, their environment, and sometimes even their personalities.

3. **Recruit mentors from the project implementation context**

   One way to ensure that needs assessment is ecologically valid and contextually sound is to recruit mentors from field sites as part of the overall project infrastructure. These mentors can provide feedback to students as they work on the projects thereby ensuring that students are on the right track and are pursuing realistic avenues in their designs. Mentors can characterize and articulate the constraints students are likely to face in their implementation. Having mentors earlier in the process is particularly important so that the feedback students receive comes earlier rather than later. Often, once design teams have reached a certain stage in their prototyping, it is hard to get them to change their ideas. Design fixation, as this condition is known in the literature, is a real challenge—students, and most designers, are reluctant to revisit their ideas if they are too deep into the process. Finally, mentors on the ground confirm the possibility for real implementation on the ground.

4. **Capture video and audio data from targeted context**

   A related step is to provide contextual information in case students are not able to experience the context first-hand. This has often been the case in projects we have undertaken. Given their international nature it is often not possible for students to be able to travel. To overcome this limitation we have developed a repository of video and audio data that the students can peruse to gain a better understanding of the context. Video in particular, given its vividness, is extremely useful in conveying an appropriate picture of the design context. In addition, we have developed a list of secondary resources that the students can go through to learn more about similar projects as well as conditions on the ground. Finally, with the use of the Internet significant information is available to students and allows them to understand the context better. All these resources work in tandem. One important lesson we have learned here is that students often lack the local knowledge to conduct useful searches on Google™ and to overcome this problem we provide them with appropriate search terms.

5. **Form interdisciplinary teams with genuine interdependence**

   One of the key insights we have gained from our experience is that the nature of problems we tackle require interdisciplinary teams. Interdisciplinary or multidisciplinary teams bring together students and faculty who approach a problem with different perspectives. Negotiations that occur around convergence of the diverse views are a true learning opportunity as well as an occasion for innovation. Design problems have a material aspect and in most cases whenever something is being designed it has an interface component and some form of hardware component. Therefore, an understanding of human interaction as well as technical know-how is required to come up with the complete solution. These projects therefore create a genuine interdependence among team members—they recognized that they needed the interdisciplinary skills to complete the project successfully. The outcomes of working in interdisciplinary teams are numerous but one of the most immediate positive outcomes is that it teaches the ability to communicate an idea across boundaries. Throughout the class sessions, the instructors emphasized the interdisciplinary expertise available within the teams as well as the need to look at the project from others' perspective to be able to form common ground.

6. **Prosocial motivation needs to exist among students**

   One of the important issues we discovered in our student focus groups, and which at some level is not very encouraging, is that prosocial motivation specific to international issues needs to exist in students for them to reap the benefits of this course. Therefore, one learning outcome from the teaching perspective is that the class should not be required of all students and will create tensions that will spoil the experience of all students if it is required. Of course, there is a dilemma here: how will students ever get motivated to take the course if they do not have exposure to global issues? We believe that many students actually have prior experiences that provide them the prosocial motivation or have done socially relevant projects in other local contexts and want to bring their experiences to a global context. We have also found that the seeds can be sowed in earlier courses through a lecture or a talk that raises students' curiosity before they work on a long term project. Sometimes students develop the motivation through other projects and societies such as "Engineers without Borders" or "Engineers for a Sustainable World."

7. **Specificity is good**

   Another key lesson we have learned is that it is absolutely fine if the projects are specific in nature as opposed to very open-ended. Although there are many benefits to open-ended projects, such as the development of cognitive flexibility, we have realized that even specific/defined projects provide students the opportunity to develop many of the skills we want them to learn from the course. They learn about developing a contextual understanding and grasping the situation from the users' perspective. They also learn the importance of working in interdisciplinary teams. Furthermore, given their unfamiliarity with the context even something that appears specific to those with prior knowledge is often open-ended from the students' perspective. Finally, the reality of projects is that they are tied to a semester or some standard time period and it is important for the students to know that their efforts within that set time period can bring useful returns.

8. **Continue projects across classes and teams**

   In terms of the execution of the projects, it is absolutely fine and even useful to continue a project across courses and allow students to tackle different aspects of the problem. The advantages of such an approach are that you can develop a useful knowledge base and it helps frame the problem in enough specificity that students can tackle it successfully. Often, students see the project in new and novel ways and this allows for even more diversity of perspective and solutions. A long term approach in terms of instruction also aligns with long-term partnerships with clients and allows incorporation of their feedback over time.

9. **Move students along the project**

   We have also realized that it is important to move students along periodically. In other words, give them weekly deadlines to keep them on track, particularly during the initial period. This is important as it is common for students to get pulled in multiple directions and not be able to reach any kind of convergence. A lack of first-hand experience with the client and design make it hard for students to trust themselves and their solutions. Therefore, they tend to linger at any one stage a lot longer than they should, and firm deadlines have to be placed to move them along to the next design stage. This does not mean that background research, needs assessment, and brainstorming should not be encouraged. It does imply that they should be limited to a certain time period beyond which students have to move along and start prototyping.

10. **Prototyping and end-product matter**

    Finally, we have realized that it is important to prototype early and prototype often. Given the goal of the course—to design a product—it is essential that students get their hands dirty. Prototyping, in addition to its advantage of demonstrating a product, also serves as a design thinking exercise and forces students to think about any design choice in different ways. It also forces them to regularly test their ideas against real-world constraints. Prototyping accelerates design thinking and also moves them along a final product that is good enough to be tested in the field or demonstrated to the client to get useful feedback.

11. **The role of intellectual influences in shaping the program**

    Finally, although we discussed many design frameworks within our class explicitly, from the instructors' and program directors' viewpoint the role of intellectual influences in shaping the program is often more tacit. We strongly believe in user-centered design, which provided the common ground to engage in a dialogue and talk about capabilities, conviviality, empowerment, and other approaches related to development. As is usually the case in collaborative partnerships, one often is more theoretically inclined than the other and this provides a balance during instruction as well as in the development of the program. Although we believe it is important for the students to "do" design, it is also important that they reflect on their experiences and are able to abstract ideas that they can apply to other problems.

12. **Context-design gap**

    One critical finding of our program has been a gap that remains between the designed artifacts and the needs on the ground due to the geographical distance between the designers and the users. It is extremely hard to capture all the idiosyncrasies of the context through any media and therefore during testing many problems with the designed artifact are uncovered. We have also realized that this diminishes learning opportunities to a very small extent but does not eliminate them. We found that students are able to learn about designing for another context and about design methodologies through this experience.

13. **Presentation of work**

Throughout our collaboration, we have worked hard on presenting our work in an attractive manner while conveying all necessary and useful information. In this, Sharma's background in design has been especially useful and industrial design students working on the project have also contributed immensely. Due to the design expertise available on the team, we have been able to design not only posters (see Figure 6.1) but presentations that demonstrate our work concisely in an engaging manner. We have realized the importance of good design through feedback and comments that we received when we presented our work but also from feedback and interest we received from our outreach partners. By being able to showcase our work in an attractive and useful manner we are able to present a highly professional image which is useful as above all it demonstrates the time and effort we put into our work. Of course, we have to be cautious about not appearing too polished and have to back up our presentation through in-depth knowledge of our work.

14. **Discussion**

We strongly believe that working on international projects related to global development not only provides students the opportunity to positively impact others but also to learn from their experiences. These projects teach students one of the most important skills they need to work in the professional world—the ability to examine something from the perspective of others. As one of our students responded, taking others' perspective was an essential but tough thing to do. Throughout this book we have been quite descriptive as we believe that learning happens from interpreting contextually rich data, as opposed to reading a list of "best practices." Therefore, although we present lessons learned at the end, we have tried to present a fair view of the overall process so that readers will be able to take away what is relevant for them. We have argued for building long-term partnerships, of carrying on projects across courses, and of working in interdisciplinary teams. We have also provided fairly descriptive data from student evaluation to show what students' themselves think they have learned. Of course we could add objective measures but we believe the outcome or products are the real measure of student success.

We want to end with a note on the need to make service learning projects, particularly long-term global/international projects, more central to the engineering curriculum. At most institutions service-learning or community-learning experiments often exist in tension with other aspects of the engineering curricula and are seen in opposition to the overall "engineering science" culture of engineering education. As Downey [2010] notes, "A main source of resistance to expanded participation in international and global engineering education is widespread acceptance by engineering faculty of the view that an engineering graduate from a given field is at the core basically one thing. That one thing is technical competence in a collection of engineering sciences" (p. 415). Similar to the marginalization of users that service learning

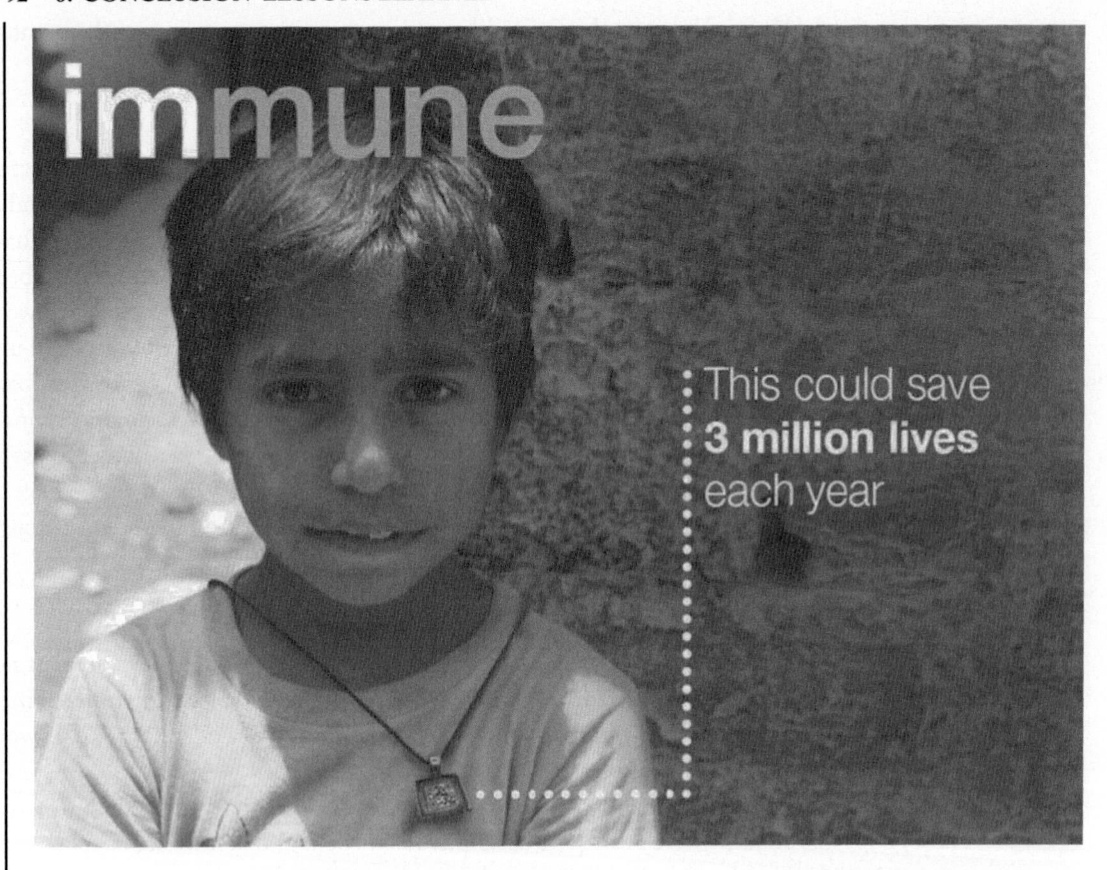

immune
This could save
3 million lives
each year

Figure 6.1: A poster advertising the project and also used as the introductory screen for the Android app.

projects aim to overcome, such projects face a similar marginalization within the curriculum particularly when they are interdisciplinary in nature and involve other liberal arts majors. One way for service-learning initiatives to become mainstream is to reject stereotypical and essentialist associations such as "girls like to help people," "racial minorities want to service poor communities," or "white males are driven by profit and high tech," and demonstrate their usefulness for all engineering and design students. Furthermore, the service learning community should recognize that intellectual exclusion, within academia, often forms the basis for social and economic exclusion, and aim at appropriating those activities that enjoy higher status such as research, publishing, award-winning, and fundraising.

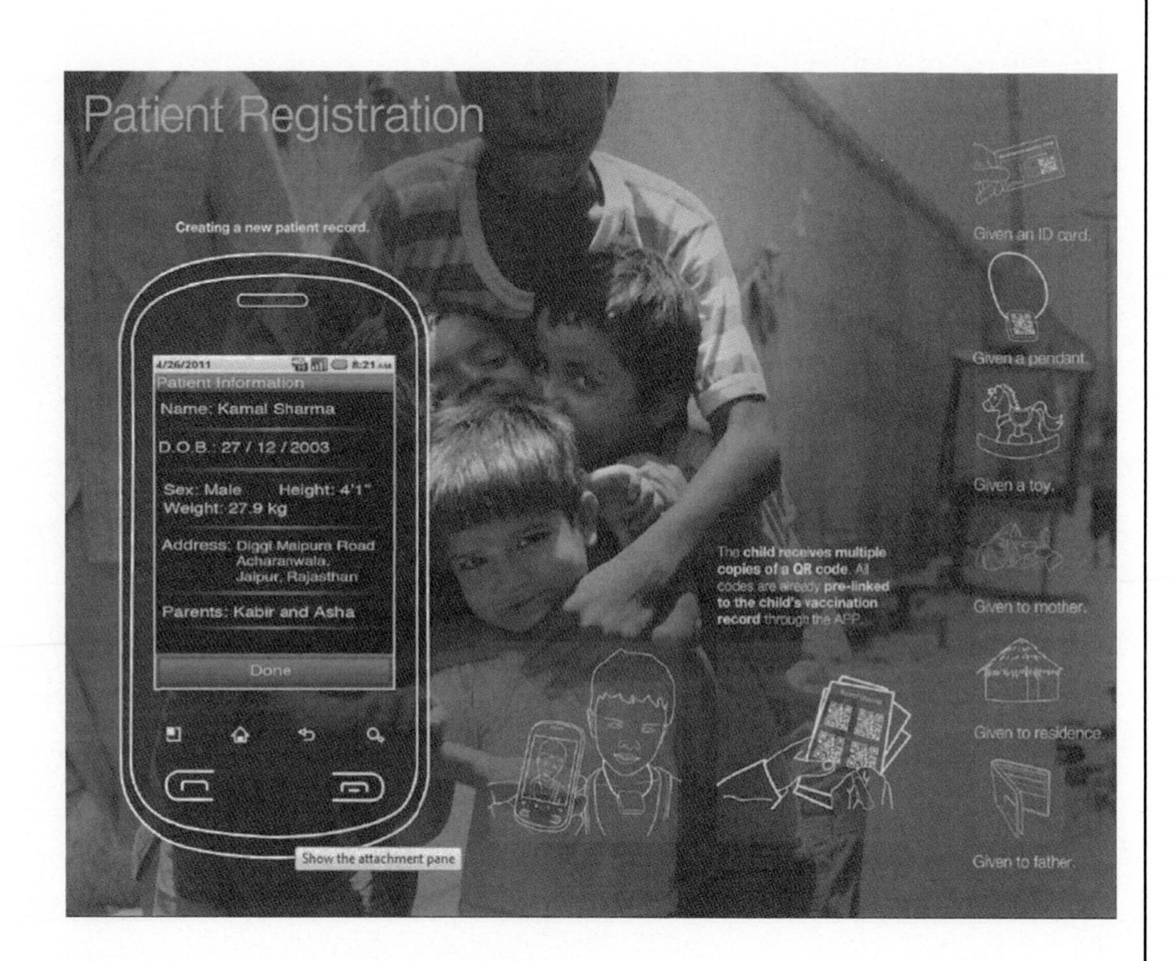

Figure 6.2: Storyboard explaining the working of the application.

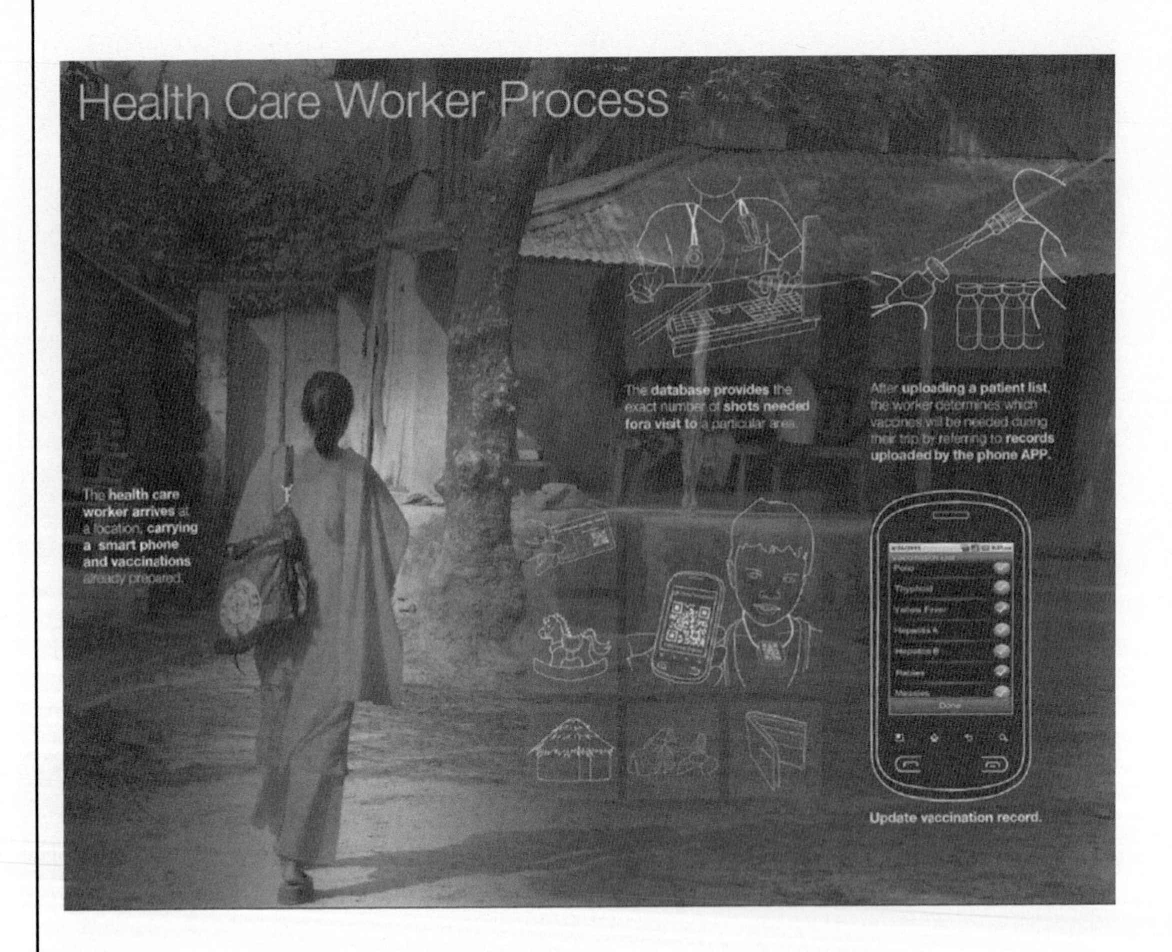

Figure 6.3: Storyboard showing the process from the perspective of the health worker.

Bibliography

Amabile, T. M. (1993). Motivational synergy: Toward new conceptualizations of intrinsic and extrinsic motivation in the workplace. *Human Resource Management Review*, 3, 185–201. DOI: 10.1016/1053-4822(93)90012-S 23

Astin, A.W., Lori J. Vogelgesang, Elaine K. Ikeda, and Jennifer A. Yee. (2000). *How Service Learning Affects Students*. Higher Education Research Institute, University of California, Los Angeles. 2

Bada, A. O. and Madon, S. (2006). Enhancing human resource development through information and communications technology. *Information Technology for Development*, 12(3), 179–183. DOI: 10.1002/itdj.20040 26

Baillie, C. (2006). *Engineers in a local and global society*, Morgan and Claypool Press, San Rafael, CA. DOI: 10.2200/S00059ED1V01Y200609ETS002 1

Barad, K. (2003). Posthumanist performativity: Toward an understanding of how matter comes to matter. *Signs*, 28(3), 801–831. DOI: 10.1086/345321 25

Batson, C. D. (1987). Prosocial motivation: Is it ever truly altruistic? In L. Berkowitz (Ed.), (1987) *Advances in experimental social psychology*, (Vol. 20, pp. 65–122). New York: Academic Press. 23

Batson, C. D. (1998). Altruism and prosocial behavior. In D. T. Gilbert, S. T. Fiske, and G. Lindzey (Eds.), *The handbook of social psychology*, (4th ed., Vol. 2, pp. 282–316). New York: McGraw-Hill. 23

Beyer, H. and Holtzblatt, K.. (1998). *Contextual design: defining customer-centered systems*. Morgan Kaufmann Publishers. 27

Bowker, G. and Star, S. L. (1999). *Sorting Things Out*. MIT Press, Cambridge, MA. 26

Bielefeldt, A.R., K.G. Paterson, and C.W. Swan. (2010). Measuring the value added from service learning in project-based engineering education. *The International Journal of Engineering Education*, 26(3): 535–546. 1

Brewer, E., Demmer, M. et al. (2005). The case for technology in developing regions. *Computer*, 38(6): 25–38. DOI: 10.1109/MC.2005.204 26

Carroll, J.M. (2000). *Making use: scenario-based design of human-computer interactions*. MIT Press. 27

Cockton, G. (2005). A Development Framework for Value-Centred Design. *Proceedings of CHI*, pp. 1292–1295. DOI: 10.1145/1056808.1056899 27

Cockton, G. (2004a). From Quality in Use to Value in the World, *CHI 2004 Extended Abstracts*, 1287–90. DOI: 10.1145/985921.986045 27

Cockton, G. (2004b). Value-Centred HCI. *Proceedings of NordiCHI 2004*, ed. A. Hyrskykari, 149–160. DOI: 10.1145/1028014.1028038 27

Coyle, E.J., Jamieson, L.H., and Oakes, W.C. (2006). Integrating Engineering Education and Community Service: Themes for the Future of Engineering Education. *Journal of Engineering Education*. January 2006, 7–11. 1

Coyle, E.J., Jamieson, L.H., and Oakes, W.C. (2005). EPICS: Engineering Projects in Community Service, *International Journal of Engineering Education*, 2005, Vol. 21, No. 1, pp. 139–150. 1

Deci, E. L., and Ryan, R. (1985). *Intrinsic motivation and self-determination in human behavior*. New York: Plenum Press. 23

Design-Based Research Collective (DBRC). (2003). Design-based research: An emerging paradigm for educational inquiry. *Educational Researcher*, 32(1), 5–8. DOI: 10.3102/0013189X032001005

Downey, G. L. (2010). Epilogue – Beyond Global Competence: Implications for Engineering Pedagogy. In Downey, G. and Beddoes, K. (eds), *What is Global Engineering Education For?*, pp. 415–432, Vol. 1. DOI: 10.2200/S00302ED1V01Y201010GES001 91

Evans, M., Johri, A., Gasson, G., Cagiltay, K., Pal, J., and Sarkar, P. (2008). ICT4D and the Learning Sciences. In the *Proceedings of International Conference of Learning Sciences 2008*. Vol. 3, pp.229–236. 9

Engeström, Y. (1999). *Activity theory and individual and social transformation. Perspectives on activity theory*, pp. 19–38, Cambridge University Press. 24

Epics (2099). EPICS, `https://engineering.purdue.edu/EPICS/`

Erickson, T. (2009). Socio-Technical Design. In Whitworth, B. and de Moor A. (Eds). *Handbook of Research on Socio-Technical Design and Social Networking Systems*. CommunitySense, The Netherlands. DOI: 10.4018/978-1-60566-264-0 25

Eyler, J.S., Dwight E.Giles, Jr., Christine M. Stenson, and Charlene J. Gray. (2001). *At A Glance: What We Know about The Effects of Service-Learning on College Students, Faculty, Institutions and Communities, 1993–2000*. Vanderbilt University. 2

Friedman, B., Kahn, P. H., Jr., and Borning, A. (2006). Value Sensitive Design and information Systems. In P. Zhang and D. Galletta (eds.), *Human-computer Interaction in Management Information Systems: Foundations*, 348–372. Armonk, New York; London, England: M.E. Sharpe. 27, 28

Friedman, B. and Kahn, P. (2003). Human Values, Ethics and Design. In *The Human Computer Interaction Handbook*, eds. J. Jacko and A. Sears, 1171–1201, Lawrence Erlbaum Associates. 27

Friedman, B. (1996). Value-sensitive Design. *interactions*, 3(6): 16–23. DOI: 10.1145/242485.242493 27

Grant, A.M. (2008). Does Intrinsic Motivation Fuel the Prosocial Fire? Motivational Synergy in Predicting Persistence, Performance, and Productivity. *Journal of Applied Psychology*, Vol. 93, No. 1, 48–58. DOI: 10.1037/0021-9010.93.1.48 23

Gurstein, M. (2003). Effective use: A community informatics strategy beyond the digital divide. *First Monday*, Vol. 8, No. 12, `http://firstmonday.org/issues/issue8_12/gurstein/` 30

Heeks, R. (2008). ICT4D 2.0: the next phase of applying ICT for International Development. *Computer*, 41(6): 26–33. DOI: 10.1109/MC.2008.192 26

Holvoet, N. (2005). The impact of microfinance on decision-making agency: evidence from south India. *Development and Change*, 36: 75–102. DOI: 10.1111/j.0012-155X.2005.00403.x

Illich, I. (1973). *Tools for Conviviality*. Perennial Library, Harpers and Row: New York. 30

Johri, A. and Pal, J. (2012). Capable and Convivial Design: A Framework for Designing Information and Communication Technology for Human Development. *Information Technology for Development*, 18(1): 61–75. DOI: 10.1080/02681102.2011.643202 28, 29, 30

Johri, A. (2011). The Sociomateriality of Learning Practices and Implications for the Field of Learning Technology. *Research in Learning Technology*, Vol. 19, Issue 3, 207–217. DOI: 10.1080/21567069.2011.624169

Johri, A. and Nair, S. (2011). The Role of Design Values in Information Systems Development for Human Benefit. *Information Technology and People*, Vol. 24, Issue 3. DOI: 10.1108/09593841111158383 27, 28

Kam, M., Ramachandran, D., Devanathan, V., Tewari, A., and Canny, J. (2007). Localized Iterative Design for Language Learning in Underdeveloped Regions: The PACE Framework. *Proceedings of ACM Conference on Human Factors in Computing Systems* (San Jose, California), April 28-May 3, 2007. DOI: 10.1145/1240624.1240791 27

Konkka, K. (2003). Indian needs—Cultural end-user research in Mumbai. In *Mobile Usability: How Nokia Changed the Face of the Mobile Phone*, eds. C. Lindholm, T. Keinonen, and M. Spencer, pp. 97–112. New York: Blackwell. 28

Khavul, S. (2010). Microfinance: Creating Opportunities for the Poor? *Academy of Management Perspectives*, 24(3): 58–72. DOI: 10.5465/AMP.2010.52842951 16

Kumaran, K. (1997). Self-help groups: an alternative to institutional credit to the poor: a case study in Andhra Pradesh. *Journal of Rural Development*, Hyderabad, India, 16: 515–530. 16

La Guardia, J. G. (2009). Developing Who I Am: A Self-Determination Theory Approach to the Establishment of Healthy Identities, *Educational Psychologist*, 44(2): 90–104. 23, 24

Latour, B. (2005). *Reassembling the social: An introduction to actor-network-theory*. Oxford: Oxford University Press. 24

Lave, J. (1987). *Cognition in practice*. New York: Cambridge University Press.

Lave, J. and Wenger, E. (1991). *Situated Learning: Legitimate Peripheral Participation*. New York, NY: Cambridge University Press. DOI: 10.1017/CBO9780511815355

Law, J. and Bijker, W. E. (1992). Postscript: Technology, stability and social theory. In W. E. Bijker and J. Law (Eds), *Shaping Technology/Building Society: Studies in Sociotechnical Change* (pp. 290–308). Cambridge, MA: MIT Press. 24

Le Dantec, C., Poole, E. and Wyche, S. (2009). Values as Lived Experience: Evolving Value Sensitive Design in Support of Value Discovery, *Proceedings of CHI*, pp. 1141–1150. DOI: 10.1145/1518701.1518875 28

Liang, L. (2010). Access Beyond Developmentalism: Technology and the Intellectual Life of the Poor. *Information Technology and International Development*, Volume 6, SE, 65–67. 30

Lieberman, H., Paterno, F., Klann, M., and Wulf, V. (2006). End-user development: An emerging paradigm. *End User Development*, 1–8, Springer. DOI: 10.1007/1-4020-5386-X_1 27

Marcus, G.E. and Saka, E. (2006). Assemblage. *Theory, Culture and Society*, 23(2–3): 1-109. 25

Marsden, G. (2008). Toward Empowered Design. *Computer*, 41(6), pp. 42–46. DOI: 10.1109/MC.2008.207 29

Morgan, D. (1997). *Focus Groups as Qualitative Research*. Sage Publications. 76

Muller, M.J. and Kuhn, S. (1993). Participatory design. *Communications of the ACM*, 36(2): 24–28. DOI: 10.1145/153571.255960 27

Mumford, E. (2000a). A socio-technical approach to systems design. *Requirements Engineering*, 5(2): 125–133. DOI: 10.1007/PL00010345 25

Mumford, E. (2000b). Socio-technical Design: An Unfiled Promise or a Future Opportunity? In *Organizational and Social Perspectives on Information Technology*, Springer. 25

Mumford, E. (2006). The story of socio-technical design: reflections in its successes, failures and potential. *Information Systems Journal* 16, 317–342. DOI: 10.1111/j.1365-2575.2006.00221.x 25

Nieusma, D. and Riley, D. (2010). Designs on development: engineering, globalization, and social justice. *Engineering Studies*, 2(1): 29–59. DOI: 10.1080/19378621003604748 1

Norris, P. (2001). *Digital Divide: Civic Engagement, Information Poverty, and the Internet*. Cambridge University Press. DOI: 10.1017/CBO9781139164887 30

Orlikowski, W.J. (2010). The sociomateriality of organizational life: Considering technology in management research. *Cambridge Journal of Economics* 34: 125–141. DOI: 10.1093/cje/bep058

Orlikowski, W.J. and Scott, S. V. (2008). Sociomateriality: Challenging the separation of technology, work and organization. *Annals of the Academy of Management* 2(1): 433–474. DOI: 10.1080/19416520802211644 25, 26

Pea, R. D. (1993). Practices of distributed intelligence and designs for education. In G. Salomon (Ed.). *Distributed Cognitions* (pp. 47–87). New York: Cambridge University Press. 24, 30

Pea, R. D. (1994). Seeing what we build together: Distributed multimedia learning environments for transformative communications. *Journal of the Learning Sciences*, 3(3): 285–299. DOI: 10.1207/s15327809jls0303_4 30

Pickering, A. (1995). *The mangle of practice*. University of Chicago Press, Chicago, IL. 24

Pipek, V. and Wulf, V. (2009). Infrastructuring: Towards an integrated perspective on the design and use of information technology. *Journal of the Association of Information Systems* (JAIS), 10(5), 306–332.

Ramachandran, D., Kam, M., Chiu, J., Canny, J., and Frankel. J. L. (2007). Social Dynamics of Early Stage Co-Design in Developing Regions. *Proceedings of ACM Conference on Human Factors in Computing Systems* (San Jose, California), April 28-May 3, 2007. DOI: 10.1145/1240624.1240790 27

Ramirez, R. (2007). Appreciating the Contribution of Broadband ICT with Rural and Remote Communities: Stepping Stones Toward an Alternative Paradigm. *The Information Society*, 23: 85–94. DOI: 10.1080/01972240701224044 29

Ryan, R. M., and Deci, E. L. (2000). Self-determination theory and the facilitation of intrinsic motivation, social development, and well-being. *American Psychologist*, 55, 68–78. DOI: 10.1037/0003-066X.55.1.68 23

Sen, A. (1999). *Development as Freedom*. Oxford University Press. 27, 28, 30

Sengers, P., Boehner, K., David, S., Kaye, J. (2005). Reflective design. *Proceedings of the 4th decennial conference on Critical computing: between sense and sensibility*, August 20–24, 2005, Aarhus, Denmark. DOI: 10.1145/1094562.1094569 27

Silva, L. and Westrup, C. (2009). Development and the promise of technological change. *Information Technology for Development*, 15(2), 59–65. DOI: 10.1002/itdj.20118 26

Star, S. L. (1999). The Ethnography of Infrastructure. *American Behavioral Scientist,* 43: 377–391. DOI: 10.1177/00027649921955326 26

Swan, C., A. Bielefeldt, and K. Paterson. (2010). Global Education: Potential Impacts of Service-Based Projects in Global Engineering Education. *World Environmental and Water Resources Congress*. May 16–20. Providence, Rhode Island. 1

UN (2000). U.N. Millennium Goals, `http://www.un.org/millenniumgoals/`

UNESCO (2008). The Global Literacy Challenge. *A Report From the United Nations Educational, Scientific and Cultural Organization.* `http://unesdoc.unesco.org/images/0016/001631/163170e.pdf`

Warschauer, M. (2002). Reconceptualizing the Digital Divide. *First Monday*, Vol. 7, No. 7. `http://firstmonday.org/issues/issue7_7/warschauer/` 30

Authors' Biographies

ADITYA JOHRI

Aditya Johri is an assistant professor in the Department of Engineering Education, Computer Science (courtesy), and Industrial and Systems Engineering (courtesy), in the College of Engineering, Virginia Tech. He received his Ph.D. from Stanford University in 2007. His research focuses on the use of information and communication technologies (ICT) for learning and knowledge sharing, with a focus on cognition in informal environments. He also examines the role of ICT in supporting distributed work among globally dispersed workers and in furthering social development in emerging economies. He received a U.S. National Science Foundation Early Career Award in 2009. He can be reached at: `ajohri@vt.edu`. More information about him is available at: `http://filebox.vt.edu/users/ajohri`

AKSHAY SHARMA

Akshay Sharma is an assistant professor in the Industrial Design Program in the School of Architecture and Design at Virginia Tech. He is passionate about using design as a catalyst for empowerment of women, especially in the developing regions. He is currently working on projects related to microfinancing with an NGO in India, use of cell phones for creating a more efficient process maintaining immunization records in developing countries, and development of a foot measurement system in conjunction with the Jaipur Foot organization in India. He received the Excellence in International Outreach Award administered through the Virginia Tech alumni association. He can be reached at: `akshay@vt.edu`.